Breaking the Science Barrier

HOW TO EXPLORE AND UNDERSTAND THE SCIENCES

Sheila Tobias
Carl T. Tomizuka

College Ent

In all of its book publishing activities the College Board endeavors to present the works of authors who are well qualified to write with authority on the subject at hand and to present accurate and timely information. However, the opinions, interpretations, and conclusions of the authors are their own and do not necessarily represent those of the College Board; nothing contained herein should be assumed to represent an official position of the College Board or any of its members.

Copies of this book are available from your local bookseller or may be ordered from College Board Publications, Box 886, New York, New York 10101-0886. The price is $14.00 plus $2.95 handling charge.

Editorial inquiries concerning this book should be directed to Editorial Office, The College Board, 45 Columbus Avenue, New York, New York 10023-6992.

Library of Congress Catalog Number: 92-071969
ISBN: 0-87447-441-8

Printed in the United States of America

9 8 7 6 5 4 3 2 1

To Mari, Frank, David, and John Tomizuka, and to Deanne Tobias Abedon, who, though not a scientist herself, produced two

Contents

3 Understanding Science 55

4 Mathematics, Models, and Measurement 99

Acknowledgments

This book benefited first from having an active scientist as one of its coauthors, someone whose training in physics, chemistry, and mathematics gave us access to a wealth of science history and fact. But no less important has been our circle of supportive friends and colleagues who were willing to suggest, amend, contribute, read, criticize, and, above all, encourage our effort to "break the science barrier" for beginners. We list them below, in no particular order; for while each brought the expertise of his or her specialty to bear on our project, it was as enthusiasts for science in general that they helped us most:

Cathy Middlecamp, Pamela Parrish, Ray Wakeland, Stephanie Singer, Rita Hoots, Ted Feragne, Mary Ellen Hunt, Stephen Abedon, Robert Hazen, Mary Kay Hemenway, Roald Hoffmann, Ruth Sime, Louise W. Knight, Diana Harris, John Spizizen, Jill Schneiderman, Michael Doyle, David Valentine, Harry Ungar, Nancy Kennedy, Joan Valentine, John Rigden, Timothy Swindle, Barbara Sawrey, Cornelius Steelink, Jennifer Schneider, Alan Van Heuvelen, and Kerry Ann Gardner, Ann Kinney, and Steve Russos.

Lastly, we thank our editor, Carolyn Trager of the College Board, and our agent, Gloria Stern, for their commitment to this project and for the many insights they contributed to our thinking about it.

Preface

This book is meant to help you discover the pleasure, power, and the lifelong usefulness of science, and thus "break the science barrier." It explores the fundamentals of understanding science and the kinds of study and thinking your professors will expect you to do. It examines the meanings of terms in science, how understanding changes from high school to college, and how you can deepen your understanding by making your own connections among facts. It describes what you should get from lectures, laboratory work, textbooks, and from the problem solving that is the "bottom line" of classroom science. Before you find yourself awash in new terms, it alerts you to the various vocabularies of science and how they are used in new concepts.

Along the way, you will meet some students just a little ahead of you in science and some professors and working scientists who are carrying on the traditions of Newton, Lavoisier, the Curies, and Darwin. We'll introduce you to a few of them and tell you about their work. Most of all, we hope to excite you about what's ahead in the study of science and to give you the information and the personal confidence needed to take the plunge.

One of the authors of this book, Carl Tomizuka, is a physicist. Trained first in Japan and then at the University of Illinois, he has specialized during most of his career in the physics of solids, studying the effects of high pressure on the properties of matter. His work borders so much on chemistry that he has collaborated regularly with physical chemists. Since 1960, when he came to the University of Arizona, he has taught introductory physics and most of the other standard courses that are required for physics majors. He has also designed and taught specialized physics courses for premedical and architectural students. In the 1980s, he developed calculus-based physics software for use with computers. Carl Tomizuka never much doubted that he would "do" science; he has a professional's command of physics and mathematics and a good working knowledge of chemistry.

Sheila Tobias, on the other hand, calls herself a "science avoider." At college, during the "bad old days" when women were not encouraged to study the physical sciences, she majored in history, literature, foreign languages, and political science. She came to science only in later years, when she discovered the extent of math and science avoidance and anxiety among college students. She has written two popular books: *Overcoming Math Anxiety* (1978) and *Succeed with Math: Every Student's Guide to Conquering Math Anxiety* (1987).

If Carl Tomizuka knows a lot about science itself, Sheila Tobias knows a lot about the barriers to learning science. She brings an awareness of what makes science seem "hard" to so many students.

1 Studying Science in College

Classroom Science

The first thing you'll probably notice when you walk into your beginning science class at college will be the size of the lecture room and the number of students there. You may feel a little overwhelmed. But it's reassuring to know that whether you're taking introductory biology, introductory physics, or general chemistry, you are about to embark on a well-worn path that many other students have traveled before and others are now traveling with you. The size of your lecture course is a tribute to how important science has become. As you will soon discover, if you haven't already, a science sequence is required for virtually all college majors except those in the fine arts and the humanities. And since college educators believe humanities and fine arts students also need to be "science literate," these students may be required to take one or more science courses to fulfill a "distribution" or breadth requirement. This is because it is more than the content of science that matters in our high-tech world. As you will soon discover in doing college-level reading, the vocabulary of science and the ways of thinking that characterize science are as much a part of manufacturing, financial management, general business, law, journalism, government, and environmental protection as they are of science itself.

Another reason for large lecture classes—one that this book

is going to help you cope with—is that college science departments find them to be an efficient way to teach large numbers of introductory students a common core of material.

As you look around, think of it this way: the students around you are as new to college science as you are. They may have taken one, two, or even three science courses in high school, but they have never before had teachers who were practicing scientists doing research at the "cutting edge" of a field of study. Although the material covered in most introductory courses, especially in physics and chemistry, is no longer on that cutting edge, college science professors often bring to the classroom the results of recent discoveries. They can't help it. It's what they are doing and thinking about, in addition to preparing for class. Your classroom association with them will give you a glimpse of what it means to search for new truths, and you'll feel a little bit of the excitement that accompanies discovery as your professor talks about what's going on in the lab and in meetings with other scientists. Even if your professor does not talk much about his or her work, you will sense how powerful it is to have enough basic knowledge to be able to frame new questions, and how thrilling it is to find answers that will stand the scrutiny of a worldwide community of scholars.

The next thing you'll notice—perhaps you will have noticed it already if you did your book buying before the first class—is the size and sheer weight of your science textbook (not to mention its cost). If you have already glanced at the course outline you may be wondering how you are ever going to digest all that is ahead. Don't worry. Much of what you will need to know to do well in this course is in the textbook and the lab manual, and the rest will be in your lecture notes. The good news about science courses is this: there is no outside reading to add to your burdens and no uncertainty about what you need to be able to do to pass the course. Your professor is your guide. Your textbook is your reference source; if you keep the book and take more courses in the field, it could become a lifelong friend.

Course Organization

Another surprise will be the amount of planning that goes into a science course. On the first day, you will be handed a syllabus containing not only a chapter outline and corresponding lecture topics by date, but a laboratory schedule (if this course has a laboratory component) and a calendar showing the homework assignments, quizzes, and exams. Your first homework assignment, perhaps a set of quantitative problems to solve or some end-of-chapter study questions to write about, may be due on the second or third day of class and your first quiz already scheduled for the second week. Partly because of the size of the class, but also because science courses have to bring you to a certain level of competence before passing you on to the next course, everything is carefully planned and scheduled in advance. Think of your syllabus as a kind of itinerary. Its detail will help you plan ahead and keep you on course—in both senses of that term.

The point is that what we call "classroom science" in this book, to distinguish it from science as it is *practiced* by scientists, is usually sequential. This means that just as each course must ready you for the one to come, many of the topics in the course you are taking will build on the one you have just learned. Science, you will rapidly discover, has to be learned in a more structured way than courses in other fields.

You will soon notice, as you start reading the textbook and attend the first few lectures, the amount and rapid pace of the work. Apart from the size and style of the lectures and demonstrations, there is the challenge of course content. In any first-year science course there are dozens of new and complex concepts to be mastered and applied, many new techniques to be learned for laboratory work and problem solving, and hundreds of new words. Someone once counted the new words in first-year biology and found there was as much new "vocabulary" as in first-year college German. Some of these words and concepts have to be memorized, but you will find that it is far wiser to study the underlying principles than to rely on simple recall.

Once you've grown used to the size and pace of your introductory courses in science, you may begin to notice more subtle kinds of differences between science and the other college courses you are taking. Most of the topics in introductory chemistry, biology, and physics, although new and perhaps difficult for you, are so basic to these fields that you won't be able to do anything original for a very long time. That's because, unlike the situation in reading courses (such as English, history, sociology, and philosophy), you don't yet know enough to make connections or to say something uniquely your own. In science, that comes with mastery and with time.

The Logic of Discovery Versus the Logic of Presentation

Initially, the subject matter of science may seem impersonal. That's because instructors need to teach beginning students what is known, logically and in terms of well-established principles. The process of discovery is always more haphazard, accidental, and fraught with wrong turns than can possibly be related in a single semester or a year-long course. For this reason the logic (or lack of logic) of discovery in science (the history of science) is replaced in the classroom by a logic of presentation.

Here's an example of how the logic of discovery contrasts with the logic of presentation. The four fundamental laws of thermodynamics (the physics of heat and its conversion) are usually presented in numerical order: the zeroth, the first, the second, and then the third. But in fact, what is known as the second law of thermodynamics (a subject you will touch on briefly in your introductory course in physics) was discovered first, in 1824, by a French physicist-engineer named Sadi Carnot. Nearly a generation later, James Prescott Joule showed experimentally that heat and motion are two manifestations of energy (the first law). William Thomson (later, Lord Kelvin) realized that Carnot's law implied two laws, the first (that energy is conserved if heat is included as energy) and the second (that heat cannot be returned to mechanical energy with 100 percent efficiency). Around the 1890s, chemists stated the third law, which continued to be revised and improved upon well into the twentieth century. Finally, what physicists call the zeroth law (having to do with temperature and equilibrium) was formulated in the 1930s as a kind of afterthought to logically round out the other three. Based on the logic of discovery, the laws of thermodynamics should be taught in the following order: second, first, third, and zeroth. However, based on the logic of presentation, the sequence is usually zeroth, first, second, and third.

As you can see, the logic of presentation may not do justice to the history of scientific discovery, but it will give you a surer grip on the subject. Science builds each concept directly on the preceding one. As a result, you will have all the knowledge you need to learn the next new concept. Confidence comes from knowing exactly what is expected of you and what you have already learned. Another reassuring feature is that classroom science, like all science, is not supposed to be arbitrary or determined by opinion. Just as every advance in science rests on experimental evidence, your performance will be based on what you do, day by day. If you learn how to work steadily at your science courses, you can expect to do well.

Science Is Dynamic

One reason for the weight of the introductory science textbook and the pace of science courses is the breadth of these subjects. Introductory courses are meant to give you the basics, but there is a tremendous volume of basics to learn. Another reason is that the sciences are dynamic, changing all the time at their frontiers. Although some new discoveries (lasers, for example) won't change the content of the introductory course, the many applications of lasers have made optics—the study of the properties of light—a "hot topic" in applied physics today. Fibre optics is crucial in telephone communication; lasers are used in surgery; and you will find a tiny laser in every compact disc player.

To take an earlier example, until 30 years ago the velocity required to escape the earth's gravitational pull (11 kilometers per second) was just a theoretical calculation. Before the era of space travel, *escape velocity* was used only to explain how we gradually lose molecules of our atmosphere. But if today's interplanetary probes are going to fly by Mars or Jupiter or photograph the rings of Saturn, they must achieve initial escape velocity to get beyond the earth's gravitational pull. When a probe goes beyond the gravitational pull of the earth, it needs to take into account the gravitational effects of the other planets as well as that of the sun. Such calculations are based entirely on Newtonian mechanics.

New discoveries do not always change science at its core. Newtonian physics is exciting because it is the conceptual foundation of all physics. It has even survived *relativity* and *quantum mechanics*—the two revolutions in modern physics—as a "special limiting case" of both (see Chapter 4). Other discoveries, such as the discovery of the role of *deoxyribonucleic acid* (*DNA*) in genetics and cell biology are, however, so far-reaching that they change the way scientists think about nearly every aspect of the biology of living things.

Genetics: Past, Present, and Future

Not very long ago, genetics was just the study of patterns of inheritance. In the past 40 years, however, there have been so many breakthroughs in our understanding of genetics and cell biology that biology has been completely transformed, and not just on the frontier. The new knowledge is so significant that even elementary biology courses have changed radically since the time your parents were in school, because of a shift in biologists' understanding of genetics and DNA.

Jennifer Schneider, now a medical doctor, can remember taking a course in genetics in the 1960s and having to mate fruit flies in her dormitory room at college. She brought home from the lab an equal number of red-eyed and white-eyed fruit flies and waited for the second generation to appear—all red-eyed. Then she mated members of the second generation with each other to produce a third generation containing both red-eyed and white-eyed fruit flies. As expected according to the laws of genetics, the ratio of red- to white-eyed fruit flies in the third generation was three to one. Schneider remembers that it was very exciting to have her work confirm what she had been told in lectures and had read in her text.

From Chromosomes to Genes

Although it was realized that there was some kind of carrier of heredity—scientists named it the *gene*—little was known at that time about the mechanisms by which genes caused red- or white-eyedness to show up in fruit flies; or how height, hair color, or any of the other plant, animal, or human traits are determined. Until the 1940s, genes were thought to be bead-like units, strung out in a line along the chromosomes, and passed on as part of the chromosomes during reproduction. Although the idea that complex living organisms have genetic material that can mutate and hence evolve was fairly well accepted by the early twentieth century, it was not yet understood that simpler microorganisms (such as bacteria and viruses) also have genes.

Scientists were hard at work to determine the biochemical nature of genes; they believed that if they understood the chemistry, and eventually the molecular structure, of genes, they would begin to understand how genes work. Since genes are located on chromosomes, an obvious first step was to determine what chemical compounds make up chromosomes. The list wasn't infinite. Biologists had known for a long time that there are only four categories of substances in living things: fats, carbohydrates (sugars and starches), proteins, and nucleic acids. Since proteins are the most chemically complex and are present in all living matter, scientists first went looking for the key genetic material among proteins. To be sure, nucleic acids, most particularly DNA, were also present in chromosomes, but, as is often the case in science, a prejudice was at work: since the transmitted traits are so varied and subtle, scientists assumed that only a very complex unit could be the transmitting agent of inheritance. Proteins were therefore thought to be the most likely candidate.

From Genes to DNA

It was only when (as a result of better measuring techniques) it became clear that there isn't the same amount of protein in every cell of an organism that scientists looked more carefully at DNA. (Protein levels vary with cell activity, but DNA does not.) In the process, they began to work with microorganisms that, whether by mutation or even evolution, can change their *phenotype* (physical appearance and behavior). But since even microorganisms have both DNA and proteins, it was still not clear which was the carrier of genetic information. Like so many questions in biology today, the answer was to come from a combination of biology and chemistry. That's why chemistry has become a corequisite of introductory biology in many colleges and why the subject of molecular biology is where the action is. But we are ahead of our story.

In the 1940s, the first big breakthrough came when a team of medical doctors began to study the bacteria that cause pneumonia, specifically the *pneumococcus* strain. What made the pneumococcus particularly interesting and made pneumonia difficult to treat was that although one strain was virulent, another was not. By carefully isolating the two strains and comparing them chemically, it was finally established beyond doubt that the "transforming substance," the part of the cell that transformed the rest, was not a protein, but was DNA.

Once DNA was identified, discoveries followed rapidly. In the 1950s, James Watson and Francis Crick, basing their findings on the experimental work of Rosalind Franklin, finally provided the full picture of DNA: a giant molecule comprising tens of thousands of atoms, of which genes are specific segments. Further work in the several new fields that grew out of these discoveries—molecular biology and genetics at the cellular level—revealed that genes control the chemical reactions that take place in living cells.

The Human Genome Project

As you learned in high school biology, but may not have understood as profoundly as you will in college science, all the cells in your body have identical genes. You can think of it this way: A DNA "library" was handed down to you, one-half from the egg and one-half from the sperm that came together at the time of your conception. After a lot of recombining, which provides unique characteristics to all living things (such as the trait of red- or white-eyedness in fruit flies), the resulting set of genes is found in every cell in your body for the rest of your life. Early in the life of a cell, by utilizing the information from a portion of these genes, it develops into a unique cell type, characteristic of a specific organ, such as a lung, a kidney, etc. The egg or sperm you produce (depending on your gender) also has a copy of this library, so some of it can be passed down to the next generation.

In a decades-long scientific undertaking called the *human genome project*, scientists are beginning the task of identifying every gene in the human body. They already know a great deal about genes. Built into the chemistry of DNA is all the information it needs, both to duplicate itself (for growth and inheritance) and to determine the functions of the cell. What is most important in these breakthroughs is that the new biochemistry of genes explains not just patterns of trait inheritance, but how living cells function. The specific organization of a segment of a DNA molecule produces certain proteins that do the chemical work of the body's cells. DNA holds the information; the cell's proteins do the work. If you doubt that this is work, consider this: as you read this page, your body is manufacturing 2,000 perfectly designed proteins, per average cell, per second.

The "New" Biology

Biology used to be mostly descriptive. The introductory course was organized by *species* and by *genus* within species, an ordering called *taxonomic* (see Chapter 2). The lectures and the textbook usually started with the simplest of living organisms and worked their way up through plants, fungi, and lower animals and finally humans. In the chapter on the squid, for example, Schneider remembers having to learn everything known about the squid: its anatomy, respiratory system, metabolism, digestive system, and reproductive system. Today's biology course is organized very differently. Given the enormous increase in knowledge of the biochemical basis of life, most biology textbooks begin with a review of some basic chemistry, and then are organized not by species but by chemical processes. The gills of a squid, in this new sequence, would be covered under "gas exchange." The chapter on how living things collect and metabolize food might be called "nutrient procurement and processing" and would discuss the whole range of food-gathering skills and digestion.

If your parents would not recognize the biology textbook you bring home from college—that's how much biology has been transformed in just a single generation—they would surely recognize the text used in introductory physics.

Physics

Unlike introductory biology, the topics covered in introductory physics have not changed much as a result of recent discoveries. That's not because the field of physics hasn't been changing; it has. It's because the basic principles—mechanics, electricity and magnetism, theories of heat, and optics—have been well understood for a long time. There may be new applications, but there are no unanswered questions of a fundamental nature. Although that makes *classical mechanics*—the first topic in introductory physics—"closed" in one sense, it remains very exciting in another sense because classical mechanics is the conceptual basis of all that follows in physics. Classical mechanics is a kind of "alphabet" of the language of physics and a "grammar" of how physicists think about the world.

Classical Mechanics

Classical mechanics is the study of the forces that cause change in motion. It was Newton who understood for the first time that what needed to be described scientifically was not motion itself, but change in an object's state of motion. Since change in the state of motion is the result of forces acting on objects, your most important task in classical mechanics will be to figure out which forces are at work on an object in a given physical situation.

Some people describe physics as a complete understanding of a very few topics. If so, classical mechanics is a good example of what makes physics uniquely "simple," and also what makes it difficult. What's simple is that there are only two kinds of forces in classical mechanics: forces exerted by contact, such as pushing and pulling real objects, and "action-at-a-distance" forces (Newton's own term), such as gravity and electricity. (Figure 1.1)

Indeed, these are the basics of physics, the concepts involved in the description of motion in space that will apply to all subsequent phenomena in physics: mechanics, electricity and magnetism, acoustics (the physics of sound), optics, and even motion inside the nucleus. More forces will be added as you proceed, such as the particular forces that bind and repel elementary particles. But the physical concepts involved in describing the motion of a simple block sliding down an inclined plane or that of a spinning top remain the same: mass, velocity, energy, momentum, and angular momentum.

Take sound, for example. Long before physicists analyzed sound, philosophers pondered whether anything would be heard

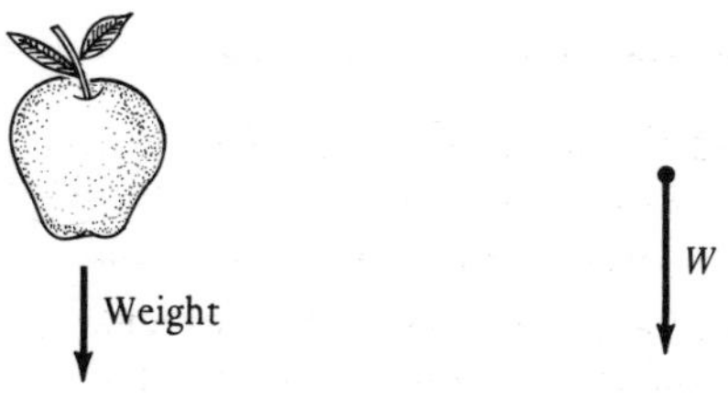

Apple falling – negligible friction

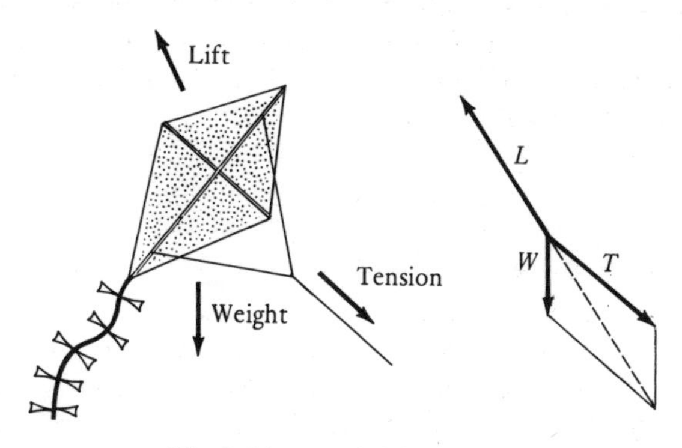

Kite held suspended in the wind

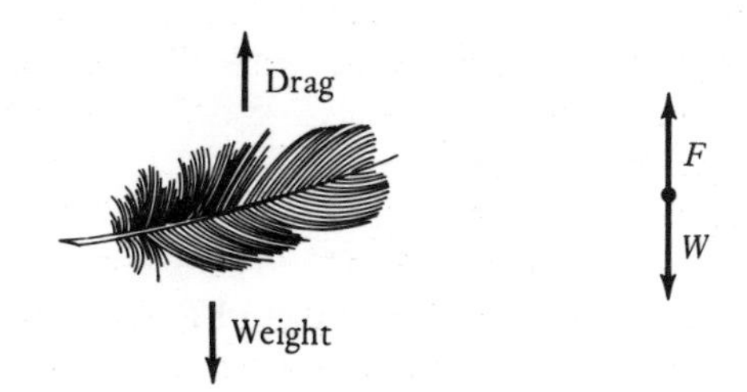

Feather falling at nearly constant speed

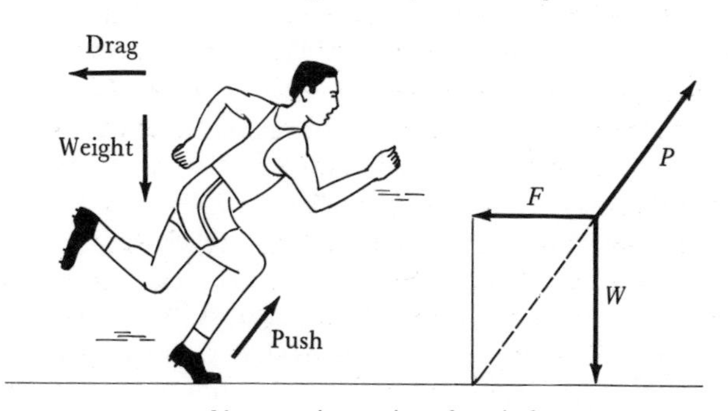

Man running against the wind

FIGURE 1.1

if no one were around to hear it. They were trying to distinguish between the physical event of sound and its perception by some sound-perceiving instrument, such as an ear. The philosopher Bishop Berkeley posed the question: "If a tree falls in the wilderness, with no one around to hear it, will it make a sound?" There was no way for early philosophers to do more than speculate about such a question, for they did not know about the physical nature of sound.

Sound may at first appear to have nothing in common with blocks moving on inclined planes, but the perception of sound is in fact the collision of particles of air with the human eardrum. The concepts you will need to understand the physics of sound are mechanical concepts: force, velocity, momentum, and pressure (force divided by area).

Waves and the Physics of Sound

Sound, as physicists describe it, is a mechanical wave motion, not too different from ocean swell (but not to be confused with ocean surf). If you look at the ocean from a distance, it looks as though the water is moving forward with the waves, but this is not the case. As Figure 1.2 shows, the water is mostly moving up and down locally. Only the energy is moving forward. You can see this quite clearly if you look from a fishing pier at a floating piece of wood or debris. Despite the appearance of a wave moving forward, that piece of wood remains pretty much in place. The same is well illustrated by the so-called "wave" of fans in a football stadium. As you know, you get up when the people around you get up, and you sit down when they do. You have not changed your seat, but to the fans on the other side of the field, the wave seems to be moving in one direction. The point of the wave is this: without that up-and-down local motion, there would be no wave. In the case of ocean swell, because the *medium* is a liquid (water), the up-and-down motion of the water is circular.

Sound waves are like water waves, but in the transmission of sound waves, the medium is air, not water. The compression and expansion of air correspond to the up-and-down local motion of water. The power of physics is to see that two phenomena as

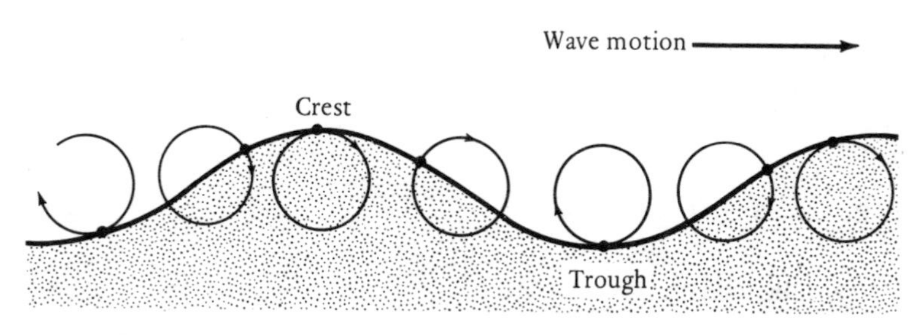

FIGURE 1.2

superficially different as sound and water waves are mechanical analogues of each other. The principles and the mathematical equation that describe the one will also describe the other.

Toward a Physics of Light

Experimentally, light has all the properties of a wave. The theory of light, completed by James Clerk Maxwell in the last century, showed that light is a moving electric field coupled to a moving magnetic field—not in a simple cause-and-effect relationship, but as inseparables, nonetheless.

But what is the medium through which light is transmitted? This question puzzled physicists for a long time. Just as the up-and-down local motion of water is necessary to transmit ocean waves and the expansion and compression of air necessary to transmit sound waves, scientists thought for a long time that there must be some *vibrating medium* in space through which light travels. It couldn't be air, since starlight travels through what appears to be the vast empty void of space. Physicists don't like exceptions to rules. Either light traveled though some medium capable of sustaining vibration but imperceptible to human senses, or light was not a wave. Since the light-was-not-a-wave theory had to be ruled out (there was too much experimental evidence), physicists went looking for a medium.

Experiments had shown that sound waves travel even faster through solids than through air. The physical reason is that atoms in solids are tightly bound to each other; atoms in gases are unattached. Physicists knew that light travels even faster than sound. Hence, logically, the medium they were looking for had to be more solid than air. Could space be filled with a solid?

The Search for "Aether"

Since light travels almost a million times faster than sound, physicists speculated that whatever medium light traveled through had to vibrate even faster, be more elastic, and be more solid than air. But space is not solid. In the 1890s, reluctant to give up concepts that so well accounted for wave motion on earth, physicists hypothesized a new medium that could not be felt or seen but had the elasticity of a solid. The hypothesis was that space was filled with "luminiferous aether" (pronounced, and often spelled, ether). Since starlight travels from very distant sources, aether must fill the entire universe. Earth must be moving through aether, too. How could this be detected?

One way was to try to measure the speed of the earth, not relative to the sun, but relative to aether. Was there an "aether wind" that the earth was moving with or against depending on the time of year? A series of careful measurements of a split beam of light was done nearly 100 years ago by two U.S. researchers, Albert Michelson and Edward Morley. Michelson (incidentally, the first U.S. winner of a Nobel Prize in physics) and Morley found no evidence for the relative motion of the earth with respect to aether.

The confusion was relieved only when Einstein later proposed his theory of special relativity.

The Uniqueness of Light

Einstein's theory of relativity asserts that, unlike any other waves, light waves do not require a medium to travel through, and that the speed of light is independent of the speed of any observer. (This explains the apparent absence of an aether wind.) There was resistance among the scientific community to Einstein's views. But eventually his theory of relativity was accepted.

This example will give you a sense of how physics proceeds

from one kind of unity to another. Physicists look for consistency and, understandably, resist giving up a theory for which there is already a large body of experimental evidence.

From Physics to Chemistry by Way of the "Big Bang"

The logic on which the "Big Bang" theory of the origin of the universe rests represents a similar kind of thinking. For some time now, astronomers have been able to estimate the speed of fast-moving sources of light in the universe. Based on these estimates, they believe that most nearby galaxies and all distant galaxies are heading away from us, and always have been. They have further observed that the more distant the galaxy, the faster it is receding; so, based on these observations, they think they know when and how the universe began: 16 billion years ago when a huge, densely packed core of matter exploded, the Big Bang. If this theory is correct, then every material known to us on earth was either present in that core or is the product of subsequent chemical and nuclear reactions. Given the laws of conservation of energy and matter, it follows that no *new* matter or energy has been created since the Big Bang. The theory is awesome in its implications. As astronomer David Helfand puts it, every atom currently in your body began in some distant star.

The Unity and Diversity of Science

Both the manufacture of proteins within living cells and the combining of elements created in the outer reaches of the universe involve chemistry. Because of this, many scientists call chemistry the central science, linking physics and biology. It is true that, except for nuclear transformations, all change in nature is the result of chemical reactions. Some (such as the manufacture of proteins in your body) occur in just microseconds. Some (such as the kinds of changes geologists study) take eons. But all are reactions among different chemical substances; these reactions are more accessible to human study today than ever before.

Basic College Chemistry

Ruth Sime, who teaches chemistry at Sacramento City College, describes the syllabus of college chemistry as follows: First, you will learn the fundamental behavior of atoms and molecules, beginning with the protons, neutrons, and electrons that are present in all atoms. This basis is derived from physics, but chemists take it from there. They are interested in the resulting differences and similarities among the hundred or so elements known on earth. For example, in chemistry you will learn just how an atom of gold differs from an atom of iron, and thus why iron rusts, but gold does not. You will also learn the arrangement of electrons in the atoms of each element—it's not too difficult because there's a pattern—and from these arrangements you will see why and how elements combine the way they do to form compounds and molecules.

You will understand why water, H_2O, always consists of two hydrogen atoms and one oxygen atom; why table salt, NaCl, has just one sodium atom for each chlorine atom; and why and how salt dissolves in water. You will learn something of the structure of elements and compounds: how, for example, carbon atoms can arrange themselves differently to form diamonds, a very hard substance, graphite, which is very soft, or the newly discovered C_{60}, which takes the form of a soccer ball. You will study how the atoms in molecules and compounds rearrange themselves as they undergo chemical reactions. Gasoline, for example, burns quickly to form CO_2 and H_2O. Sugar burns slowly in your body to form the same substances. Both are chemical reactions producing usable energy.

Further Specialization in Science

Chemistry is large in scope. In your first course you will study the basics of composition, structure, and reactions—the essential framework for the more specialized courses that lie ahead, such as organic chemistry, analytical chemistry, physical chemistry, biochemistry, and the many applications of chemistry in the biomedical, agricultural, environmental, and engineering sciences.

After completing the introductory course in biology, you can go on to plant or animal anatomy and physiology, microbiology (the biology of microorganisms), ecology (how organisms interact with their environment), and evolutionary biology. Still more advanced are courses in genetics, population biology, and molecular biology.

After the introductory physics courses in mechanics and electricity and magnetism, if you decide to major in physics, you will go on to thermodynamics (the physics of heat), optics (the physics of light), advanced mechanics, electrodynamics, and relativity; these, in turn, can open the door to quantum mechanics and particle and nuclear physics. And of course, with a strong grounding in any one of the physical sciences—not necessarily a major—you can build a career in engineering.

Why Not One Science of Nature?

Why, you may wonder, are there three basic sciences at all? Why not just one science of nature? The answer lies in the changeover from natural philosophy to experimental science during the modern period. For the past two hundred years, chemists have studied substances; physicists, forces and motion; and biologists, living organisms. Only in this century has it become clear that at the molecular level biology depends on a knowledge of chemistry, and that at the atomic level chemistry depends on a knowledge of physics. But the fields still don't merge because the interests of their practitioners are not fundamentally the same. The chemist wants to study molecular activity and is interested in creating new molecules. The biologist is interested not only in the chemistry of cells and parts of cells but in their functions (and ultimately in the question: what is life?). And when physicists study solids and liquids, they are not interested in their chemistry so much as in their physical properties, such as electrical conductivity.

So, although you are well advised to take introductory courses in all three sciences, at some point you will need to decide which questions and which answers—the chemist's, the physicist's, or the biologist's—suit you best. Many students who finish a year or two of college science are surprised that they enjoyed these courses so much. This is understandable. College science is not just what happens in the classroom; it is also your first glimpse of what scientists actually do.

Who Should Study Science?

As you think about whether you want to pursue science in college, some people who specialize in the study of population trends are thinking about you. They are concerned with what they call science and engineering *demographics*; whether there will be enough young people in the United States preparing to become scientists, technicians, and engineers to provide the needed workers. Based on current figures (four million working scientists and engineers, of whom approximately two-thirds are employed in industry), some are worried about a possible shortfall in science and engineering by 2010.

One fact stands out: of the total number of Ph.D.'s in science and engineering now being awarded annually, only 41 percent are earned by U.S. citizens. Although some foreign students may decide to stay, many others will return to their home countries, taking their technical expertise with them. To counteract this loss, more U.S. citizens should be majoring in science. Also students majoring in other fields should be taking more elective science courses to improve their science literacy as citizens.

Another significant statistic is that in the current work force (115.5 million people), 47 percent are white males, 36 percent are white females, and 17 percent are nonwhites of both sexes. By 2010, when there are expected to be 25 million more people working than now, nearly two-thirds of the work force will be made up of the groups (including women) that have not traditionally chosen science as a career.[1]

What these statistics mean for you is this: Especially if you are a female or a member of a minority group, you are likely to be heavily recruited and enthusiastically welcomed if you major

1. "Final Report of the Task Force on Women, Minorities and the Handicapped in Science and Engineering," (National Science Foundation, 1988).

in science. Many universities are already providing tuition scholarships in science, mathematics, and engineering; summer internships; on-campus research jobs; and summer science-related travel. We have to assume that if the nation needs to attract and keep professionals in science, there will, in the long run, be very good benefits and pay.

Are You a "Science Type"?

Are you happiest working alone? Do you work well with other people? Although there are still "loners" in science—particularly among those who specialize in theory—most experimental scientists work in teams. Laboratory work makes for fine friends and colleagues. Whether in a university, a research institute, or in private industry, the lab is the place where senior researchers share tips and insights with their postdoctoral assistants and trainees. The goals of the science lab are serious, and the work standards are high. Above all, there's the sheer pleasure that comes from working closely with minds as good as or better than your own.

Do you like foreign languages? It is certainly true that English has replaced German as the language of science, and mathematics remains the international code, so you could probably succeed without knowledge of other languages. But science is such an international enterprise that the scientist or engineer who knows Russian or Japanese, for example, has a wider selection of collaborators and can be counted on to know about important new findings, wherever they occur, without waiting for the information to be translated.

Do you like travel? There are thousands of scientific meetings worldwide in any one year, and tens of thousands of U.S. scientists and engineers are traveling to them. One friend of ours, a structural engineer, went to scientific meetings in Turkey, Brazil, and China in just one 18-month period.

Are you a nonconformist who can't imagine yourself getting suited or dressed up every morning? Science (except in certain industrial labs and at formal gatherings) is as fashion-blind as it should be color- and gender-blind and is tolerant of all but the most bizarre personal styles. There was a graduate student in microbiology at the University of Arizona who, after seeing the movie "The Last Emperor," shaved half his head in the style of the young Emperor of China; no one in his microbiology lab objected or even seemed to notice.

Do you like working with your hands or mostly with your mind? Creating complicated gadgets or working at the chalkboard? Puzzling out solutions to problems or finding new problems to work on? Do you like being indoors or out? Is writing something you really enjoy or would you rather hire someone to do it for you? Do you like to talk? Teach? Think? Draw? Hack at the computer? There's something in science for every kind of style and temperament, so long as you have the basic requirements: intelligence, persistence, patience, open-mindedness, and creative imagination. The reason? There is so much science left to do!

Many people earn more money than scientists do, but few seem to enjoy their work more. Jonas Salk, the inventor of the polio vaccine, is still working beyond the age of 70, long after he might have retired. He is now putting all his efforts into understanding the AIDS virus. Linus Pauling, who did important research on the nature of the chemical bond as well as on the DNA molecule, just celebrated his ninetieth birthday, but still works regularly in his lab. How would you like to make a living doing what you enjoy? The true comparison is not between science and business; it is between science and the arts. For in their own way, scientists are artists. For many people, the appeal of lucrative occupations is that one can afford to retire early. The appeal of science and the arts is that you won't want to!

Gaining a Competitive Edge

But suppose you don't want to major in science. Suppose your career, at least at this stage of your thinking, doesn't even require a lab science for graduation. Should you consider taking introductory chemistry, biology, physics, mathematics, or computer science anyway? Consider this possible future scenario: You're competing for a reporter's job on a newspaper, the dream job you've been pursuing since you were the editor of your high school paper. In a kind of preinterview competition, the editor distributes some news and feature assignments randomly. A third of those stories are medical and pharmaceutical, or have to do with wildlife ecology, toxic waste, earthquakes, agriculture, air and water quality, or water resources. You don't know everything that bears on each of these news stories, but you know enough chemistry and biology and physics to find out more. Your editor is pleased; your colleagues are impressed; the job is yours.

There are many such possibilities. Perhaps you love working in the theater or TV, but you're not interested in performing. How long will it take you to learn the computer light board, electronic video editing, or sound engineering if you have no familiarity with physics, computer science, and mathematics? Or, suppose you become a counselor for the emotionally disturbed and you have patients who are being treated concurrently with mood-altering medication. Will you be comfortable consulting with the resident psychiatrist about chemical interactions, dosages, and side effects?

Recently, there was a truck spill on a busy San Diego freeway. Police came, cordoned off the freeway for six hours, and evacuated many thousands of people living in the area. The spill, as it turned out, was iron oxide. But the people who responded to the call didn't know enough chemistry to take responsibility for deciding that iron oxide is just rust and could have been handled at much less cost and disruption by merely hosing down the chemical.

Basic knowledge of chemistry, biology, physics, and computer science and mathematics will make you science literate in a world that is becoming ever more technically sophisticated. Without such a background will you be able to

- Assess different solutions to environmental degradation problems, genetic engineering, or health cost containment?
- Take a position on some new weapon system and hold it in the face of opposition?
- Do the necessary legal research as a lawyer if your company decided to go into the field of superconductivity?
- Protect yourself and your family if radon or some other toxic agent were suddenly discovered in your basement?

These and many other pressing problems require a basic knowledge of science. Let's hope you say "Yes!" to science.

2

The Vocabularies of Science

A myth—as exaggerated and misleading as most myths are—is common among precollege students: one is either "good in English" or "good in math and science," but rarely in both. The fact is that scientists and mathematicians usually also score high on the verbal sections of standardized tests. In their professional work, scientists have to be fluent in both verbal language and the universal languages of symbols and mathematics. What is different about learning science is that it may sometimes be more efficient to think in symbols than in words and to write down equations rather than sentences. One college student discovered, when looking over her physics notes in preparation for an exam, that she hadn't written down any words at all in her notebook. The pages were filled with equations, diagrams, drawings, symbols, numbers, and graphs. Science cannot be learned without language skills. However, you won't learn science by memorizing vocabulary; you will acquire the vocabulary by learning science. If you approach the vocabularies of science the way you study new words in other subjects, you will fall far short of understanding. The student who comes to college used to studying by memorizing terms may have some adjusting to do.

What is true for science is true for all the specialized subject matter you will be studying in college. Most of the new words you will have to learn are technical terms, words that have a particular, narrow meaning, specific to that subject. Scientists, like all specialists, want a standard language that is unambiguous, one with which they can communicate precisely what they mean. Since science rests on verifiability (experiments have to be repeatable and reproducible), clarity, and precision, the use of standard terms is vital. But this is not unique to science. The term *Zeitgeist* in the writings of the German philosopher Hegel is not just the literal translation "time-spirit," any more than democracy is fully explained by the definition "government of the people." These words represent concepts so embedded in the volumes of interpretation that have been written about them that no dictionary definition could possibly do them justice. (In fact, if you look them up in an encyclopedia, instead of in a dictionary, you will find whole essays written about them.)

What makes the terms in science uniquely challenging for beginning students is that they are new, there are so many of them, and, in addition to words, you have to master each subject's symbol systems. There are abbreviations, such as pH and NMR, in chemistry, and ATP in biology; chemical formulas, such as NaCl; mathematical expressions, such as $F = -kx$ (the force law of a spring) in physics; terms with Latin or Greek roots, such as *atom* (in Greek, "uncuttable") and *molecule* (in Latin, "little heap"); "process" words, such as *oxidation* and *reduction*; and units and/or rules and laws named after their discoverers, such as *Haber cycle* in chemistry, *joule* in physics, and *E. coli* in biology.

The vocabularies of science call for fine distinctions not always needed in spoken English (for example, the distinction between *solvent* and *solute*). So however skilled you are at your own language, English, you are going to have to learn additional and quite different ways of communicating in science. The good news is that once you master a term in science, there's little room for error. You will know exactly what is meant when a term is used. Your instructor will know exactly what you mean when you use it. And as soon as you become fluent in these vocabularies, you will be able to talk to everyone else working in a particular field of science, whatever their native languages happen to be. The vocabulary of mathematics, which is mainly symbolic, is particularly universal, as mathematician Miriam Leiva found out when she emigrated from Spanish-speaking Cuba in the 1950s.

Miriam Leiva was 14 years old when her parents left their native Cuba for the United States. When she began to attend ninth grade here, she knew no English, nor were her teachers and classmates able to translate her subjects for her into Spanish. Dutifully she attended five of her six classes on her first day of school, not understanding a word of either what the teachers said or what was on the chalkboards. Finally, she walked into her sixth period class, Algebra I, and there, as she tells the story, "Everything was in Spanish!" The algebraic symbols on the board and the equations in the textbook were exactly the same as the ones she had already learned in Cuba. The universality of mathematical notation had made this displaced teenager feel at home. Not surprisingly, Leiva became a professional mathematician and is now a professor of mathematics at the University of North Carolina at Charlotte.

Many terms in science are just as universal as mathematical expressions. They are the same in all languages and do not have to be translated. In fact, to maintain standardization, certain international scientific associations meet periodically to approve new notation and to arbitrate differences in existing terminology so that the various symbol systems will remain intelligible to all.

USING TERMS

Except at the very beginning, you will not usually be asked in college science to define a term. You show that you understand a term by using it correctly in, for example, balancing an equation in chemistry, solving a motion problem in physics, or identifying a biological process. It is in the doing that the learning takes place in science. For example, a textbook in beginning chemistry states:

> A buffer solution must contain an acid to react with any OH^- ions that may be added to it, and a base to react with any added H^+ ions. Furthermore, the acid and the base components of the buffer must not consume each other in a neutralization reaction. These requirements are satisfied by an acid-base conjugate pair (a weak acid and its conjugate base or a weak base and its conjugate acid).[1]

1. Raymond Chang, *Chemistry* 3rd edition, (New York, Random House, 1988), 673.

This is clearly written in the vocabulary of chemistry. Indeed, the author is generous in giving the reader the meaning of the term *acid-base conjugate pair*. To understand it, you will need to know the meanings of *buffer solution, neutralization reactions*, and *acid-base conjugate pair* as used in chemistry. Except for the connective words, you might say this sentence is written in code, but it is a special kind of code. Knowing the bare-bones definitions of these terms—in the dictionary sense—won't tell you much unless you already know the chemistry involved. On the other hand, if you have gotten this far in general chemistry (the quoted material appears in Chapter l6 of the textbook), each of these terms should speak volumes to you. You will have done experiments in the laboratory with buffer solutions. You will have measured the changes and lack of changes in the pH (a measure of base or acidity) for many different solutions, and you will know how buffer solutions are constructed. So, by Chapter 16, the text will not be a mystery, but rather a summary of ideas and concepts you have already experienced in the lab.

This is what science students mean when they tell us: "You don't get into science by learning the words. You get into the words by learning science."

Artificial Languages

From a linguist's point of view, the vocabularies of science are artificial, not natural (like those in spoken English and other spoken languages). That means they are deliberately constructed, and new terms are always being added. As new phenomena are discovered and old phenomena are better understood, the vocabularies of science are constantly in flux, growing and expanding. There was, for example, no term for the *nucleus* of an atom (not to be confused with the nucleus of a cell) before scientists in the twentieth century discovered that atoms have central cores. Earlier, scientists imagined the atom to be something like a raisin pudding; the "pudding" was a glob of positive charge, and the "raisins" were negatively charged electrons embedded in the glob. But around 1911 or 1912, three physicists—Ernest Rutherford, Hans Geiger, and Geiger's student, Ernest Marsden—discovered that instead of a pudding, there was a small but very dense positively charged core at the center of the atom, with electrons distributed around it. Now that there was a core, that core had to be named, and Rutherford chose the name nucleus.[2] When, 35 years later, it was discovered that fast-moving neutrons could split the nucleus of uranium atoms, the German physicist Lise Meitner had to come up with a term that would describe atomic splitting. Her choice was *fission* (from biology), and we are all familiar with that term and its implications.

The same evolution of terminology has occurred in chemistry. The first chemists recognized only the metals—gold, lead, iron, copper, and silver—as elementary substances. In the eighteenth century, when chemists found that certain gases—the ones we call today oxygen, hydrogen, and nitrogen—were also elements, these needed to be named. Although science was then, as it always has been, an international enterprise, there were not yet any international agreements as to language, so the gases took on different names in French, English, and German. Lavoisier's term *oxygen* works just as well in English as it does in the original French (*oxygene*) because of its Latin root. But the German language isn't based on Latin. Since chemists thought (wrongly) at the time that oxygen was the essential ingredient in acids, the Germans named the new element *Sauerstoff* (acid stuff), and the name has been retained. Hydrogen, not surprisingly, became *Wasserstoff* (water stuff) and nitrogen, *Stickstoff* (suffocating stuff), because in pure nitrogen gas many living things suffocate. If the Germans were naming hydrogen today, they might call it *Kosmosstoff* because hydrogen, as the astronomer Cecelia Payne-Gaposchkin discovered only 60 years ago, is the most plentiful element in the universe.

As more and more new elements were discovered, scientists reached further for their names, sometimes after the place where the element was first discovered, such as germanium for Germany, scandium for Scandinavia, and helium because it was found in the sun (Greek: *helis*); polonium was named by one of its discov-

2. The term nucleus had a long history in other scientific contexts, but not in physics. Biologists thought of the nucleus as the innermost part of a living cell; astronomers used it for the brightest part of a comet's head. Once nucleus was decided on for the central core of the atom, scientists in other countries translated it into their own languages. Hence, the word for nucleus in German is *kern* (meaning "pit" as in a cherry pit). The Japanese word is *kaku*, which also means "core."

erers, Marie Curie, after her native Poland. Radium was named from the French word *radioemanation* because it radiated "invisible rays." When two more laboratory-newcomers to the periodic table (see Chapter 3) were discovered in the 1940s, they were named neptunium (number 93) and plutonium (number 94) because, like the planets Neptune and Pluto whose orbits lie outside that of Uranus, they are beyond—in the sense of their *atomic numbers*—uranium, number 92.

As these examples illustrate, the vocabularies of science can be said to evolve as understanding changes. And just as biological evolution eliminates certain species and favors others, so some terms in science die out when the ideas on which they were based are found to be wrong. Such a term was *caloric,* coined when eighteenth century physicists and chemists wrongly thought heat to be an indestructible fluid substance. Caloric, as they used it, was a kind of river of heat that flowed from a hot object to a cold one, as from the top of a mountain to the bottom. Once a truer model of heat was understood, one that links heat with motion (of molecules), the term caloric, as used in this way, had to be discarded (although the unit of energy in the form of heat is still called "calorie," and is used to refer to the caloric value of foods). Sometimes terms are kept on even after their original meaning no longer applies. We now know that the atom is not an indivisible particle of matter, as the Greeks thought it to be. But we still keep the term because atom has taken on a meaning of its own.

One might say that discarded scientific terms provide a good corrective to the all-too-common view that scientists are never wrong. In fact, "wrong" ideas are common, but all ideas are constantly being tested, so that science can move forward.

What is not altogether efficient in science is that the vocabularies, the notations, and even some of the symbol systems are not uniform across fields; sometimes the same words are used to convey different ideas even within the same fields. Chemical nomenclature, it is said, has a lot of "potholes"; at least half a dozen variables/constants in chemistry are called K or k. Even though chemistry and physics share a theory of heat, the quantity known as F in physics is very frequently symbolized by A in chemistry. Atomic and molecular masses in physics are often referred to as atomic and molecular weights in chemistry. Meanwhile, biologists measure the molecular mass of DNA in terms of *base pairs* and the mass of proteins by the number of their amino acids. So be wary as you cross fields, for you may be crossing language boundaries, too.

The Origins of Terms

Terms in science, as we have already seen, may be taken from a number of sources. The most obvious is everyday language and experience. Terms such as work, force, chaos, and relativity in physics come from ordinary speech, as do salt, equivalence, compound, and bond in chemistry and cell, respiration, and fermentation in biology. These terms are all borrowed, so to speak, from our English vocabulary. But in each case, the term is given a more precise meaning when it is used in science. At other times, common words are given an altogether different meaning in science. A *normal force* in physics is the force at right angles to the surface. In computer science, the word *default* means to do something in the way previously specified, and not, as in common English, failure to pay a bill.

Another source of terms in science is the names of the scientists who made discoveries in a particular field. For example, Escherichia coli, commonly called E. coli, is a particular species of bacteria, named for the bacteriologist Theodor Escherich; a joule is a unit of energy, named for the physicist James Prescott Joule; the *Hardy-Weinberg Principle* is named for the mathematician, G.H. Hardy, and the physician, Wilhelm Weinberg, who independently discovered this particular law of population genetics. Sometimes, a term is named after a person who may not have done the key calculation or who may even have been altogether wrong. An example of the first case is *Avogadro's number*. Amadeo Avogadro, an Italian physicist who worked in the nineteenth century, never actually calculated the number, because the means to do that calculation, such as X-ray crystallography or Millikan's experiment to determine the elementary electric charge, did not become available until a century later. But he did have the critical insight; that at the same temperature and pressure, the same gas volumes will always contain the same number of gas molecules—whatever the gas. So when a constant, based on his theory, was finally determined to be 6.02×10^{23}, the law and the number were appropriately named after Avogadro. An example of the second case is the *Zener diode*, named for Clarence Zener's *electrical breakdown theory*. Subsequent research has shown that this particular electrical breakdown is not accounted for by Zener's theory, but the name has been retained.

Today, it is less likely that a term will be named after a single scientist. The fact is that most scientific discoveries are made by more than one person, a team of coworkers, or even two teams working independently and coming to the same discovery at about the same time. How to name these team efforts? Sometimes a group of researchers will link the first initials of their several last names together. So, for example, *TAPPS* stands for the five scientists who developed the theory of "nuclear winter," and the *WAY* theory in physics is the way scientists remember that Eugene Wigner, Fujihiro Araki, and Michael Yanase did the work.

What happens when a scientific term is named after a living person? Does he or she use the term like everyone else? As in all matters, this varies by individuals. The late John Bardeen, who received one Nobel Prize for the invention of the transistor and another for his theory of superconductivity, never liked to call the superconductivity theory by its official name, *B-C-S*, after himself (B) and his two colleagues Cooper (C) and Schrieffer (S). Instead, he would refer to it, modestly, as *microscopic theory*. On the other hand, the late Italian physicist Enrico Fermi, sometimes considered the inventor of the atom bomb and not known for his modesty, used to refer to *Fermi distributions* with a grin. Chemists, as a rule, tend to be more modest than physicists. The discoverers of the *Ziegler-Natta* catalyst, for which another Nobel Prize was awarded, call it by its lengthy chemical name rather than by their own.

We have already seen that scientists can be playful as well as serious in selecting new terms. Rooted though they are in reality, they don't mind using their imaginations, or someone else's. The *boojum*, an odd-looking desert plant, was named after a creature Lewis Carroll created in his fantasy, *Alice in Wonderland*. Although the *Southern blot* in molecular biology was named after its founder, a biologist named Southern, similar techniques, devised later, were whimsically named *Western blot* and *Northern blot* to designate the different substrates used. In the same fashion, computer scientists like to break down *bytes* into *nibbles*. The *quark*, a subnuclear particle that scientists now believe to be the ultimate constituent part of neutrons and protons, comes from the James Joyce novel, *Finnegan's Wake*, a fantasy about life in Ireland. Physicist Murray

Gell-Mann picked the term *quark*, even though it meant something altogether different in Joyce's novel: a "quart" (of ale or beer).

The Vocabulary of Mathematics

As Leiva discovered at age 14, mathematics is the universal "shorthand" to express quantitative relations in nature. More than that, it is the most efficient way to describe what is going on. If a is related to b by a constant (k), then the mathematical expression, $a = \mathrm{k}b$ is a concise and unambiguous way to express a is proportional to b with a proportionality constant k. As relationships become more complex, mathematical expressions become the only way to express them, and these mathematical expressions often become the source of insight and predictive power that scientists are looking for in their work and that words simply cannot supply. For this reason physical scientists in particular consider mathematics to be the framework for thinking in science.

As you will quickly learn in college, mathematics involves much more than mere computation. Mathematics helps construct *models* of natural phenomena that scientists use both to take measurements and to make predictions. (See Chapter 4.) This is what Galileo meant when he wrote, four hundred years ago, "The Book of Nature is written in the language of mathematics." Professor David Layzer, who teaches astronomy and physics at Harvard University, likes to begin his physics class with the following comparison: "A physics course without mathematics is like a vegetarian beef stew. It may be appetizing and nutritious, but it doesn't resemble the real thing." Albert Einstein went so far as to speculate that God might have been a mathematician. Because the description of nature is so mathematical, the relation between math and science goes both ways. Just as courses in math ready you for courses in science, so science gives you lots of practice in mathematics.

Some students, however, do not find mathematics empowering. It is for them just the reverse, a barrier to learning science.[3]

3. If you think you have this problem, see Sheila Tobias, *Succeed with Math: Every Student's Guide to Conquering Math Anxiety*, (New York, The College Board, 1987).

They are afraid their knowledge of mathematics is too weak to serve them in science. But successful students of science, even those for whom mathematics did not come easily in high school, tell us that science clears up many of the mysteries of mathematics, because in science mathematics is presented in terms of concrete examples.

Algebraic Expressions

Let's look at a few examples to get a feeling for the usefulness of mathematics in science. The law of universal gravitation, expressed in words, states that:

> Every material object attracts every other material object with a force directly proportional to the product of the masses of the two objects, and inversely proportional to the square of the distance between them.

This sounds complicated. Using a mathematical expression, however, this law can be expressed succinctly as:

$$F = G \frac{m_1 m_2}{R^2}$$

> where F is the force of attraction, G is the universal gravitational constant, m_1 and m_2 the objects (masses), and R is the distance between them.

Especially in chemistry and physics, one of the most important habits beginning students need to acquire is thinking algebraically—thinking about relationships among quantities rather than about the numerical quantities themselves. The following information, for example, is given in quantitative language, but it is not nearly as useful as the algebraic generalization that follows:

> If one drops a stone from a height of 10 meters, it strikes the ground at a speed of 14 meters per second.

This can be expressed algebraically by the following equation:

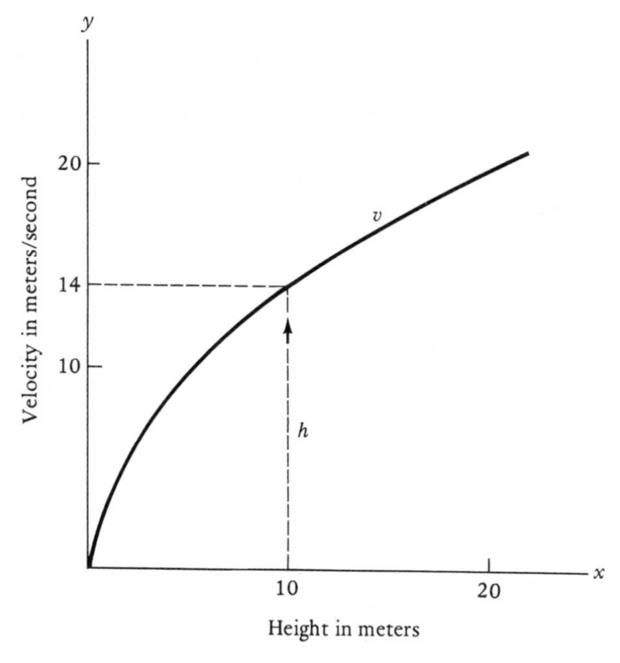

FIGURE 2.1

$$v^2 = 2\ gh$$

where v is the velocity, g is the acceleration due to gravity, and h is the height.

The reason the equation is so much more useful than the numerical words is this: algebra displays *functional relations*. If you can understand the algebra, you will be able to grasp how all the individual quantities affect all the others (see Figure 2.1); in this equation, how a change in the height (h) will affect the velocity (v) with which the object strikes the ground. Another way of saying this is that the algebraic expression describes the relationships over the entire graph; the numerical approach gives you information at only one point (↑) on the graph.

The algebra lets you see that g can be a variable, too. On the surface of the moon or on Jupiter, where gravity (g) is different from that on earth, you can still employ the same mathematical expression to describe a stone falling to the surface by simply changing the value of g.

Despite the clarity of an algebraic equation, you must still read the accompanying text carefully to learn the conditions under which a particular equation is valid.

Graphing Relationships

Because scientists see relationships mathematically, it always helps to plot functional relations on a graph. This doesn't mean you always need to work on graph paper. Sometimes graphs need to be done precisely; other times, they are meant to be just sketches, not even drawn to scale. The *slope of the curve* will give you a good sense of what is going on in a particular motion problem or in a chemical reaction. The *area under the curve* will give you a measure (or a prediction) of the total amount of change that has gone on in some chemical, physical, or biological process.

Plotting relations on a graph, therefore, is as common in science as writing out equations. Visualization is so important in the way scientists think about relationships that some scientific calculators come with a graphics capability that can be activated at the push of a button.

The Vocabulary of Biology

The terms you will have to learn in biology are the most numerous but, in many ways, the most straightforward. They are, for the most part, names of species and processes; there are also names of organisms, parts of organisms, and biological molecules. Biology has the largest vocabulary for one simple reason: there are so many unique entities in the world of living things. Chemistry deals with unique substances, too. But since these substances are made up of a few more than the 92 naturally occurring elements, chemical formulas are constructed out of just over 92 "letters" of a chemical alphabet. Once you know the elements and the principles of combining, you can determine the meaning of the whole.

What's in a Name?

The system of grouping organisms in biology, called the kingdoms of life, was invented by Carl von Linne, a Swedish scientist who lived and worked during the eighteenth century. Von Linne chose Latin for his classification system and even changed his own name to its Latin equivalent—Carolus Linnaeus. Linnaeus assigned each known organism to a kingdom, a phylum (or division), a class, an order, a family, a genus and a species. The last two, the genus and the species combined, are used as a sort of "proper name" of an organism, its two-name *binomial.* As in all naming systems, there are rules in constructing the binomial. The first name is the genus and is capitalized; the second is the species and is not. Thus, *Homo sapiens* in the Linnean system means genus *Homo*, species *sapiens* (thinking). (Although they have long been extinct, there were other species of *Homo* before modern humans: *Homo erectus* and *Homo habilis.*)

Linnaeus did not merely name the organisms; he created a hierarchy of classifications based on key aspects of their structure. His kingdoms, phyla, classes, orders, families, genera (the plural of genus), and species are related in a discernible manner. Remember that microscopes in the eighteenth century were primitive, so the only differences scientists could deal with were those that were very marked and visible. For example, organisms were classified according to whether they had fins, tails, gills, etc., and then subdivided according to other obvious characteristics. Humans and flies belong to the same Linnean kingdom (Animalia), but not to the same phylum. Just as we are *Homo sapiens*, the laboratory species of fruit fly is *Drosophila* (genus fruit fly) *melanogaster* (specific species).

Living organisms are still classified as plants, animals, and fungi. But with today's instrumentation, finer distinctions can be observed on the microscopic level. Among microorganisms, biologists can now distinguish single- from multicelled creatures. At the cellular level, they have discovered some single-celled species with cells containing a fully developed nucleus (membranes that enclose their DNA); they named these species *eukaryotes.* Other species, whose cells are not "nucleated" (the DNA is not enclosed but is drifting in the cells), have been named *prokaryotes.* This discovery added two more kingdoms to Linnaeus's list.

The effort to learn the names of these binomials and to recognize the species they refer to will be a lot easier if you pay attention to prefixes, suffixes, and word roots. "Pos," for example, means foot; "ped" means child (but it can mean foot, too, as in pedal); "karyote" means nucleus; and "eu" means true or good. Hence, a eukaryote is a truly nucleated organism. Since prokaryotes are believed to have evolved earlier, they are given the prefix "pro" (sometimes "pre"). Even with such clues, you had better be prepared to memorize a lot of new terms in biology; they anchor ideas.

New Classification Systems

With the development of new knowledge in biology, the old classification systems are being enlarged and refined. One widely used textbook in introductory college biology[4] recommends a five-kingdom classification system, which divides multicellular organisms into plants, animals, and fungi, and unicellular organisms into eukaryotes and prokaryotes. An example of how new knowledge influences classification is the case of *bacteria*. They used to be listed under plants, although it was obvious that bacteria are in no way like ordinary plants. When it became possible to study bacteria under better microscopes, it became clear that bacteria are motile (able to travel under their own power). That means they are not plants but should be classified as animals in the Linnean system.

Although bacteria now have a place in biologists' classifications, viruses are in a kind of limbo. The fact is that scientists cannot agree on whether viruses are alive. It depends on how you define "alive." Most living things pass on their genetic material *and* produce proteins. Viruses do the former, but not the latter. They pass on their genetic material, but have no cell membranes and no machinery for making proteins. A virus "survives" by exploiting its host cell's machinery. Away from the host cell, it behaves chemically like a nonliving molecule. So is it living or not? You'll notice that this is not just a scientific argument, but a philosophical one as well.

4. Morris Keeton, *Biological Science*, (New York, W.W. Norton Co., 1987).

The "virus problem" gives us a sense of how scientists in general, and biologists in particular, try to cope with the vast amount of information they have to think about. On the one hand, they must have some kind of system of classification to organize the material. On the other hand, new knowledge breaks down old systems. In biology, there is another problem: nature is not as orderly as scientists would like. Many differentiations overlap; many species or subspecies are "missing." This is because evolution itself was not orderly, but rather the consequence of local solutions to local problems. Hence the results of evolution are something of a hodgepodge and the terms in biology are many and varied. No wonder students compare its vocabulary to that of a foreign language.

The Vocabulary of Chemistry

The vocabulary of chemistry is more than a naming system. It is a bridge between thinking in words and thinking in chemistry itself. There are nearly seven million known chemical compounds, and that number increases by about 10,000 every year. Clearly, there is no way anyone can know the names of all these compounds, both natural and synthetic. But there are other terms, symbols, process words, and names of chemical quantities that give beginning students a basis for learning chemistry. Some of these will be discussed in the pages that follow.

If you took a chemistry course in high school, you know that there are different kinds of terms in chemistry. There are the terms for microscopic entities, such as atom, molecule, bond, and ion, which are components of the atomic/molecular *model* of matter that chemists and physicists employ. (For more on models in science, see Chapter 4.) Hence no textbook definition can do them justice. Only when you become familiar with the model, will you really understand the words.

Another vocabulary in chemistry consists of the names and formulas of specific substances. These are usually abbreviated for convenience as combinations of single- and two-letter symbols. Some elements are hydrogen (H), sodium (Na), chlorine (Cl), oxygen (O), potassium (K), and uranium (U); some compounds are sodium chloride ($NaCl$), water (H_2O), and ammonia (NH_3).

Once you master the chemical naming system, you will find yourself becoming chemically "literate"—able to think in chemical code.

A third vocabulary consists of process terms (which define chemical changes), such as reaction, oxidation, and reduction; and quantity terms, such as *molarity*, *solubility*, and the *mole*. (See Chapter 3.) Your textbook will give you some standard definitions of these terms to get you started, but real understanding comes with use. In English, you might have a definition wrong and not know it until your teacher grades your work. In science in general and in chemistry in particular, the terms are tightly tied to the facts. If you understand a term, you will get positive feedback in working the problem; if you confuse two terms, you won't be able to solve the problem. That's why Catherine Middlecamp, who tutors general chemistry at the University of Wisconsin, says that successful students of chemistry do more than just memorize terms. They know that textbook definitions are only the starting point.

Chemical Formulas

Just as we attach names to people, chemists attach names and corresponding formulas to substances. The chemical composition is created by nature; the formulas are deciphered by chemists. Your professor makes an immediate connection between a chemical formula, such as CH_3OH, and its chemical name, methanol. When you can do this for the many substances you will be introduced to in general chemistry, you will be well on your way.

Simple chemical formulas have become so much a part of our everyday language that most of us know that H_2O represents water; NaCl, table salt; and H_2SO_4, sulfuric acid. The formula is standard chemical shorthand to describe the atomic contents of the molecule. To a chemist, H_2O means that two hydrogen atoms and one oxygen atom combine to make one water molecule; NaCl means that one positively charged sodium ion (ions are positively or negatively charged atoms) combines with one (negatively charged) chloride ion to produce table salt; H_2SO_4 means that two hydrogen atoms combine with one sulfur atom and four oxygen atoms to form sulfuric acid. This much is fairly straight-

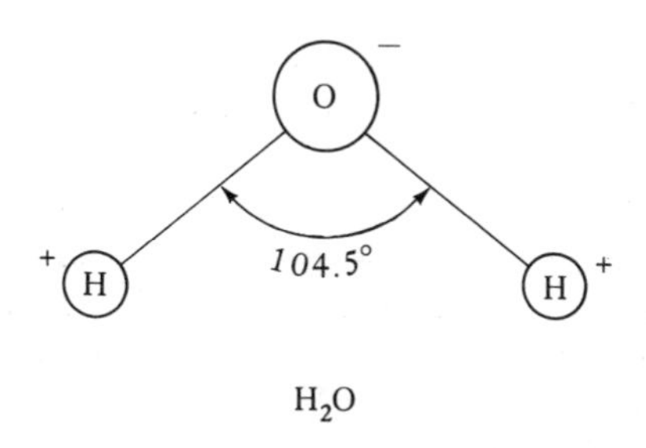

FIGURE 2.2
The spatial layout of the water molecule.

forward. However, these terms can have subtly different meanings, depending on their context.

For example, H_2O can mean any of the following to a chemist: a single molecule of water; any amount of water (a raindrop, a potful, or a reservoir); and one mole of water—18 grams. (See Chapter 3 for more about the mole.) This may seem confusing, even imprecise, but not to a chemist. Since all water molecules are identical, the number of them present in a raindrop, in a reservoir, or in one mole doesn't change the essence of water. The same molecules, each consisting of two hydrogen atoms and one oxygen atom, are present no matter what the quantity. Focusing on essentials, the chemist expresses that sameness with the formula H_2O. Thus, every chemical formula can have a microscopic meaning (one molecule), a macroscopic meaning (any amount of water), and a quantitative usage (one mole).

As we saw in the chemical formulas above, the subscripts tell us how many atoms there are in a particular molecule but don't provide any clues as to the geometry of the compound. The formula H_2O tells us only that two hydrogen atoms are joined (bonded) to one oxygen atom, but not how—in what spatial layout. Yet, the geometry of the molecule is one of its most important features. The lines between the hydrogen atoms and the oxygen atom in Figure 2.2 represent the *chemical bond*, and the angle between the two bonds (the *bond angle*) measures 104.0 degrees.

CH_3OH is the formula for methanol (wood alcohol, the kind that is toxic to humans). Ethanol (grain alcohol, the kind that makes you tipsy and is also toxic to humans in high doses), has

Methyl Alcohol: CH_3OH Ethyl Alcohol: CH_3CH_2OH

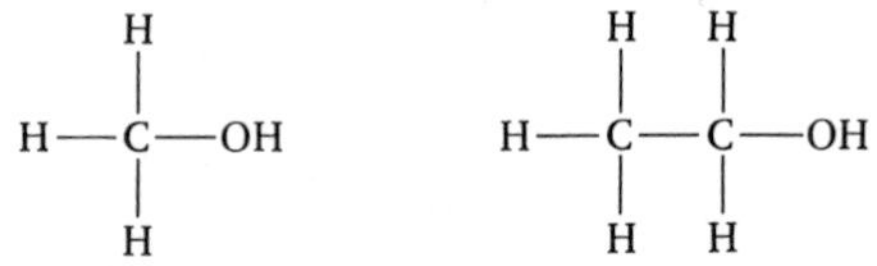

```
      H               H   H
      |               |   |
  H---C---OH      H---C---C---OH
      |               |   |
      H               H   H
```

FIGURE 2.3
Position of atoms in alcohol molecules.

the formula CH_3CH_2OH, just one CH_2 group more than methanol. The subscripts in the formula for methanol tell us that there is one carbon atom for every three hydrogen atoms, and one more hydrogen atom for the oxygen atom. Why don't chemists write the formula rather as CH_4O, combining the four hydrogens? (Or C_2H_6O in the case of ethanol?) The reason has to do with the geometry of the molecule; chemical formulas show not only the number and kinds of atoms that make up compounds, but try to indicate how they are connected to each other. Methanol is a combination of CH_3 and OH, and ethanol is a combination of CH_3,CH_2, and OH (Figure 2.3). Here each "C" represents a carbon atom, each "H" represents a hydrogen atom, and each "OH" represents an oxygen and hydrogen pair.

For chemists, the chemical formula summarizes some important information about the molecule: which atoms are present, how many there are of each, and how they are connected. Some information is not given: the details of chemical bonding, and how electrons are shared. But this is not a problem once you learn the logic of chemical bonding. As Middlecamp says, the more you know in chemistry, the more you see in the formulas. By the time you get to CH_3OH, you won't have any trouble visualizing the likely structure of this molecule. You will have enough information about bonding to figure it out for yourself.

Chemical Reaction Equations

In addition to chemical formulas, your course in college chemistry will teach you to "read" *chemical reaction equations* which describe how and in what ratios substances react.

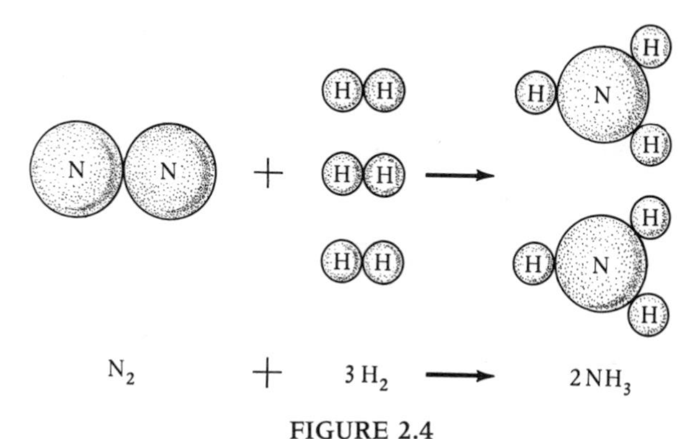

FIGURE 2.4
Position of atoms in ammonium molecule.

Reactants	Product
$N_2 + 3H_2 \longrightarrow$	$2NH_3$

is a reaction equation that describes what happens when one nitrogen molecule combines with three hydrogen molecules. The result is two new molecules, each consisting of one nitrogen atom and three hydrogen atoms (Figure 2.4). Note that the nitrogen and hydrogen molecules at the beginning of the reaction have two atoms each. At the end of the reaction, the atoms have been rearranged. This is typical of chemical reactions. Here each "N" represents a nitrogen atom and each "H" represents a hydrogen atom.

The key information in a chemical reaction equation is the whole numbers that precede the chemical formulas. These are the *combining ratios*, or what chemists call the *stoichiometric coefficients*. In Figure 2.4, the combining ratios are: 1 nitrogen (if no number appears, it is understood to be 1), 3 hydrogen, 2 NH_3. They represent the ratio of the number of molecules that combine and the number of molecules that result in the product. Note that although 4 molecules rearrange to form 2 new molecules, the numbers of N atoms and H atoms on each side of the reaction are the same. That's why this is a balanced chemical reaction equation.[5]

5. Instead of an "equals" sign between them, chemists use an arrow $\longrightarrow$ to indicate the direction of the reaction. A double arrow $\longleftrightarrow$ means the reaction is in chemical equilibrium, that is, changes in both directions are taking place, and the rates of reaction to the right and to the left are the same.

Acid Rain and Ozone: Chemical Reactions in the Environment

Consider the following reaction:

Reactants		Product
$SO_2 + O_3 + H_2O$	$\longrightarrow$	$H_2SO_4 + O_2$

It means that sulfur dioxide (SO_2)—the gas given off by some industrial processes—reacts with water (H_2O) and ozone (O_3) in the atmosphere to produce sulfuric acid, which returns to earth as a component of the *acid rain* that is ruining our landscape and defacing our buildings. (Nitric acid is also present in acid rain, the result of a similar reaction with the nitrogen in the atmosphere, induced by sunlight.) Because acid rain releases toxic chemicals from the soil that then flow into water sources, it is causing a decline in the population of fish in our northern lakes. It's also killing our trees. Since weather knows no boundaries, acid rain falls on areas that do not produce sulfuric acid as well as on areas that do, souring relations among nations and regions. Thus it has become an international political issue as well as an environmental one. Anyone who has studied enough chemistry to understand the reactions that produce sulfur dioxide and nitric acid will have a technical edge over people who haven't and, more important, much insight into how and how not to try to solve the problem.

Many chemical reactions in natural surroundings occur in series. If you've ever wondered why environmentalists oppose the use of aerosols and refrigerants, the answer can be found in the series of chemical reactions that threaten the ozone (O_3) that is found in the upper atmosphere. This ozone absorbs much of the sun's ultraviolet rays, which cause skin cancer. Freon is the brand name of two organic molecules in which some or all of the hydrogen atoms are replaced by fluorine and chlorine atoms. (The chemical name for compounds like these is *chlorofluorocarbons*, or *CFCs*.) When exposed to sunlight, these substances undergo *photodecomposition* (decomposition caused by light), which releases chlorine. The following reaction series then takes place:

Reaction 1

Reactants		Product
$Cl + O_3$	$\longrightarrow$	$ClO + O_2$

Reaction 2

$$\begin{array}{lcl} \text{Reactants} & & \text{Product} \\ ClO + O & \longrightarrow & Cl + O_2 \end{array}$$

In Reaction 1, one atom of chlorine (Cl) reacts with one molecule of ozone (O_3) to produce a chlorine-oxygen molecule (ClO) and an oxygen molecule (O_2). In the course of the reaction, the valuable ozone molecule is broken up. Once it loses an oxygen atom, it ceases to be ozone (O_3); instead, we have ClO and oxygen (O_2). (See the right-hand side of the equation.) As if this were not damaging enough, in Reaction 2 the new ClO molecule immediately starts reacting with any free oxygen atom, producing another chlorine atom (Cl) and oxygen (O_2). The newly formed chlorine is then free to attack another ozone molecule, and so on. What's so damaging about this series of reactions is that they start with a chlorine atom (Reaction 1) and end with another chlorine atom (Reaction 2). In Pac-Man-like fashion, the chlorine gobbles up ozone again and again, destroying many thousands of ozone molecules.

Atmospheric chemists did not foresee the dangers to the ozone layer in the use of Freon in consumer products. It was not until sometime in the 1970s that scientists began to notice that, at certain times of the year, the ozone layer over the Antarctic disappeared. The only way to eliminate the chlorine-ozone reaction is to ban CFCs or to find ways to recycle them; as the reaction sequence described above demonstrates, it will be a long time before all the chlorine in the atmosphere is eliminated. The United States government recently agreed to ban new products that emit CFCs, which is a beginning.

Diagrammatic Representations

As we have seen in Figures 2.2, 2.3, and 2.4, every chemical entity occupies three-dimensional space, and its spatial architecture determines some of its most important properties, including how it reacts with other molecules. Thus, it could be argued that chemistry's symbols are not really vocabulary, but rather a bridge between pictures and words. The simplest, indeed a much oversimplified drawing of a molecule, is in terms of balls and sticks (Figure 2.5).

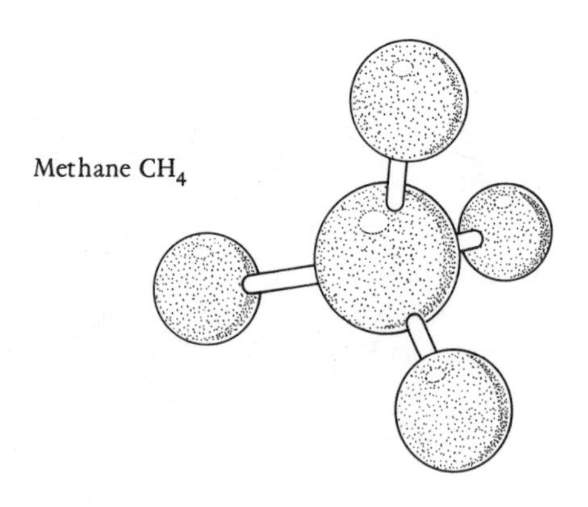

FIGURE 2.5
Methane CH_4 molecule. The large "ball" is a carbon atom; the four smaller balls are hydrogen atoms.

Such pictures are helpful in thinking about what happens at the microscopic level where atoms join to form molecules. But you won't see many ball-and-stick models in college chemistry because they show only the fact that atoms bond, not how they bond, that is, how they share electrons. As you learn more about bonding, you will note that the diagrams will represent different degrees of detail. For instance, the *Lewis* or *dot* diagrams shown in Figure 2.6 indicate which electron pairs are shared and which are available to be shared in any given molecule. But, as in all two-dimensional representations, there is always a trade-off. In

```
      H                H
      |                ..
  H — C — H        H  :C:  H        Methane—CH4
      |                ..
      H                H

      ..               ..
  H — O:           H  :O:           Water—H2O
      |                ..
      H                H

      ..               ..
  H — N — H        H  :N:  H        Ammonia—NH3
      |                ..
      H                H
```

FIGURE 2.6
Lewis or dot diagrams

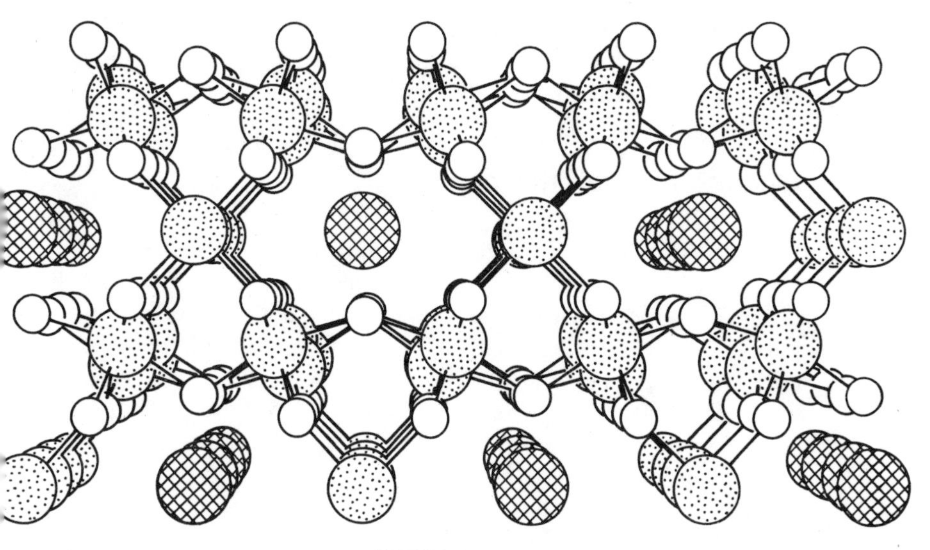

FIGURE 2.7
Hoffmann drawing

this case, the actual bond angles aren't usually 90 degrees, and the molecules certainly aren't flat.

Words, pictures, and symbols are equally important to chemists. Roald Hoffmann, who won the 1982 Nobel prize in chemistry, tells us that in a typical 20-page research paper that he submits to a chemistry journal there will be as many as 90 drawings, many of which he produces himself (an example is shown in Figure 2.7). He finds these extremely useful in explaining his work to others, but he includes them also because he finds much beauty in chemical architecture. Learning to think visually in chemistry will help you remember what you learn and begin to appreciate what chemists see when they think about matter.

The Vocabulary of Physics

There is good news and bad news about the vocabulary of physics. On the good side, there are fewer new terms to learn in physics than in the other sciences. On the bad side, those terms must be *operationally defined,* that is, defined in terms of how and what is measured. There are only four basic quantities physicists need to describe nature: length, mass, time, and electric *charge*. All other physical quantities are derived from these four basic quantities and can be expressed as combinations of units of them.

Operational Definitions

Take speed, or what physicists prefer to call *velocity*, for example.[6] The velocity of a moving object is expressed in physics as a distance covered during a certain time interval, divided by that time interval. Operationally, distance is defined as the difference between two readings of a ruler, and time interval is defined as the difference between two readings of a clock. So if the reading on your odometer is 120 miles and the reading on your watch is two hours, your average speed is 60 miles per hour.

This may seem a roundabout way of defining something that is simple and straightforward. The advantage of the exactitude with which physicists approach measurement shows up as seen in the definition of a more complicated concept: *acceleration*. Physicists define acceleration more specifically than we do. In physics, acceleration has a broader meaning than the common usage—speeding up. It is used to mean any nonuniform motion, whether speeding up, slowing down, or changing direction. For physicists, acceleration is the *change in velocity divided by the time interval in which the change took place*. You can see how this definition builds directly on the definition of velocity. Velocity is length (displacement, i.e., change in position) divided by time; acceleration is velocity divided once again by time. To become airborne, a typical commercial airliner increases its ground speed from 0 miles per hour to approximately 180 miles per hour in about 30 seconds. Assuming that its engine thrust is constant, its average acceleration (change in velocity over the time interval in which the change takes place) is about 6 miles per hour, per second. Physicists prefer that all measurements be made in standard units, so they would describe the airplane's acceleration as a certain number of meters per second, every second; the symbolic expression is m/s/s, or

6. Velocity is defined in physics as a *vector* quantity, one that has direction as well as magnitude. Speed, as used more colloquially, is a *scalar* quantity, one that has only magnitude.

m/s^2. For this airplane example, 6 miles per hour per second corresponds to about 2.7 m/s^2.

Note how crucial to these definitions is the idea of quantity. Before Galileo, Descartes, and Newton, the physical world was thought about in terms of "essences," "purposes," and fixed causes. The great strides in physics came about when these vague concepts were replaced by quantifiables. For physicists, a definition that is not operational, that is, one that does not include exact instructions as to what to measure and how to measure it, is useless.

Perhaps the best illustration of the nature and power of operational definitions is given by two terms—*mass* and *force*—that you will be working with in beginning physics. The dictionary defines mass as "a quantity of matter, forming a body of indefinite shape and size; a lump." A physicist will not be satisfied with a definition like that, because without knowing how to measure it, there is no meaning in physics for mass.

Isaac Newton (1642-1727) is revered by physicists as the codiscoverer of calculus, the founder of mechanics, and the discoverer of the law of universal gravitation. He understood the importance of finding ways to measure physical quantities. His attempt to define mass, although not successful, shows that he was intuitively groping for a definition that would give him something specific to measure. He defined mass as a "quantity of matter . . . measured by its density and bulk." Historians of science are not certain what Newton meant by "density" or how he intended to measure it. (We measure density as mass divided by volume, so that wouldn't have given him the definition he wanted.) But no matter. His instinct was correct.

Physicists got around the density-volume problem when they began thinking about mass as an object in a state of motion. Today, mass is viewed as a measure of the *inertia* (resistance to motion) of a body. The more mass a body has, the "harder" it is to change its speed or its direction, i.e., its *state of motion*. But what does harder mean? And how do we measure harder? (Now we are thinking the way physicists do.) Physicists use harder to mean how much more force would be required to move an object from a state of rest, or to change its direction of motion. To define a "measure of inertia," we have to define a "measure of force." Since force is defined as a cause for the change in the state of motion and mass as a measure of inertia, the resistance to that change, the two terms have to be defined simultaneously by means of Newton's second law of motion: force equals mass times acceleration ($F = ma$). Often in physics you start looking for a law and you end up with a definition, or you start looking for a definition, and you end up with a law. That's because laws and definitions are interconnected. When you study the laws of physics, it will help if you ask yourself each time: How much of this is definition? How much of this has to do with a law of nature?

Physicists are interested in quantities not as mere numbers, but rather in relation to other quantities. A velocity of 60 miles per hour or an acceleration of 6 miles per hour per second is not as interesting or useful to a physicist as is a general symbolic expression, such as $F = ma$, that links several variables. Since algebra is the mathematics of variables and their relations, and calculus is the mathematics of change, both are the natural form in which physical quantities are expressed. Mathematics is the language of physics.

When physicists began defining physical phenomena operationally, they were able to leap beyond the ancients' views to more accurate and deeper understandings of how we describe nature. Operationalism implies an operator; measuring involves instruments and human beings to read the measurements. How do we correct for the limitations of human experience? The ancients got into trouble when they relied on unprocessed sensory data. That's why they thought objects on earth were naturally at rest, and objects in the heavens were naturally circumnavigating the earth. When they realized that human beings are part of a moving universe, that is, when they began to imagine what they could not experience, Copernicus, Galileo, Kepler, and Newton gave us a more accurate depiction of the physical world.

After Faraday and Maxwell, and due in part to their work, there was another leap forward in thinking about physical laws early in the twentieth century. Albert Einstein made us realize that the measurements we take of natural phenomena may vary depending on where we stand and how we are moving. You know from your experience of traveling at high speeds in an automobile that as you drive by, the trees lining the road appear to be moving as fast as you are, but in the opposite direction. Einstein said that if your speed approached that of the speed of light, those trees would appear not only to be moving, but to be narrowing in width. In a post-Einsteinian universe, the operational definition

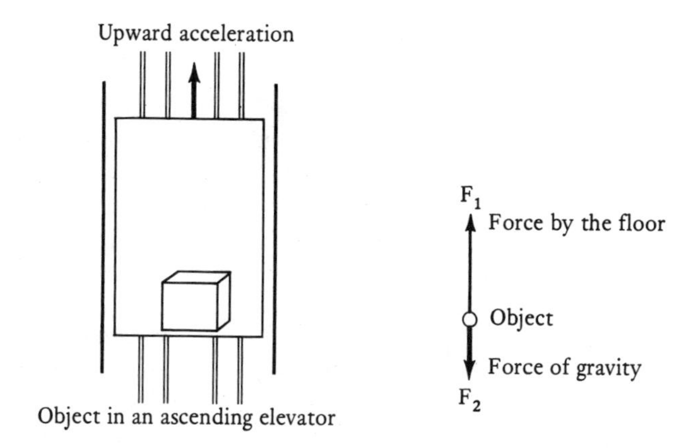

FIGURE 2.8

Free-body diagram of the object in the elevator as described by an observer on the ground (outside the elevator).

of any physical quantity (including length and width) involves the *relative motion* of the observer to the observed.

Frames of Reference

You won't have to deal with Einstein's theory of relativity right away in physics, but it is useful to know that your instructor's way of thinking and describing motion is much influenced by relativity. That's why you will be asked to take into account the position and state of motion of an observer. This is thinking in terms of a specific *frame of reference*. It will be a lot easier to analyze problems such as that of an object falling in an ascending elevator if you distinguish the observer's frame of reference. Is the observer traveling with the elevator, or standing outside it? Once you select the frame of reference in which to describe motion, you are then ready to think about forces acting on various parts of the system you are looking at, that is, constructing the all-important *free-body diagram* (Figure 2.8) which is so useful in solving physics problems.

The force on the block by the floor (F_1) is upward, and the force on the block due to gravity (F_2) is downward. Since F_1 is larger than F_2, the elevator is accelerating upwards.

The vocabulary of physics is relatively easy for students to master because in terms of *kinematics* (the description of motion without regard to the cause of motion with which your physics course will begin), it makes no difference what kind of object is moving—blocks, monkeys, cars, or stars—it's all the same in physics. That's why you won't have nearly as much verbal detail to master as in biology and chemistry. What makes physics difficult is what makes physics interesting. Once you accept that basic terms such as time and length have to be defined operationally and in terms of some specific frame of reference, you will find yourself confronting some of the deepest questions that have led to advances in physics. It is said of Einstein that his thought explorations into relativity began with "simple" questions, such as "What time is it? How do I know? How do I measure it?" Until then, people had thought of time as something that flowed uniformly everywhere in the universe. After Einstein, they knew that time is relativistic: to a stationary observer, a clock traveling close to the speed of light will appear to slow down.

The vocabulary of physics, then, is closely linked to what physicists do and how they think about the world.

Measurement—Units and Standards

No chapter on the vocabularies of science would be complete without some discussion of measurement, and of the units and measurement standards that scientists employ. Units are the given names of certain quantities, such as gallons, liters, or yards; standards are the physical entities on which these units are based. The standardization of measures and the creation of a system of universal units is an important part of human history. From earliest times, people employed units of measurement—most of them based on anatomical dimensions, the daily rotation of the earth, and other observable regularities of nature. We still have the remnants of some of these: the English and American yard, for instance, was originally a measure of the distance from a person's (any person's) nose to the fingertip of an outstretched arm. If you put your nose against your shoulder, you can see that it is a good way to measure cloth from a bolt of material. You pull the material as far as the end of your arm and then you cut it. But given the vagaries of reach, the yard, as it was originally defined, could hardly have been a universal, exact measure across boundaries.

This problem was as obvious to the ancients as it is to us. The Egyptians tried to solve it as early as 3000 B.C. They started with the *cubit*, standardized as the length of some royal person's arm from the elbow to the outstretched fingertips. They called this the royal cubit, and created a block of black granite in which this length was etched. (Thus, the cubic was their measure, and the block of black granite their standard.) All cubits used in Egypt from that time onward had to be measured against this royal master cubit, which created some standardization of length. So while their measure of length was not nearly as exact as ours today, they were on the right track.

The Babylonians, Hebrews, and other ancient peoples struggled with the same task. Archaeologists have found remains of Babylonian measures for length, mass, and volume. More important, the Babylonians exported their units and standards to other peoples in the Mediterranean world. The Greeks needed linear and area measurements both for their impressive architecture and for their sports competitions. They were the first to link the "finger" with the "foot." The Romans made five feet equal to one pace, or double step, a thousand of which made up the Roman mile. What the Greeks and Romans were attempting was not just standardization, but connection between units.

Still, for centuries there were variations. One standard length would be used in carpentry and another entirely different one in tailoring. The system of weights and measures we have inherited from the British, although eventually standardized with the best means at their disposal, is still a hodgepodge, compared to the metric system. Except for our currency, which is decimal, we are not accustomed to thinking in magnitudes of 10s. We have feet, inches, and yards, measured in divisions of 12 and 3; acres of 640 square yards; three kinds of ounces, liquid volume (16 ounces in a pint), avoirdupois (16 ounces in a pound), and troy ounces (12 to a troy pound and used for weighing precious metals). In England, there is still another weighing measure—the stone (roughly 14 pounds). Because of this inheritance, U.S. students of science have to overcome old habits in order to begin to use what is now a worldwide standard, the metric system of measures, based on 10.

The Metric System

You may not have learned this in your world history course, but one of the most significant results of the French Revolution was the establishment of the metric system of weights and measures. The idea was contained in a prerevolutionary proposal to create an entirely new system based on decimalization, rational prefixes (such as deci-, centi-, milli-, kilo-, and mega-),[7] and use of the earth's measurement as the basis for standardization. In 1790, in the midst of all the political upheavals, the French Academy of Sciences determined that the length of the meridian passing through Paris be measured exactly from the North Pole to the Equator, and that one ten-millionth of that length be called one meter. It also ordered that a new measure of mass, one gram, be set at the mass of one cubic centimeter of water (at its temperature of maximum density, 4 degrees Celsius or 39.2 degrees Fahrenheit). A list of prefixes, the ones still in use, was formulated, and the whole proposal was passed by the new National Assembly.

By the time the king was supposed to authorize the measuring of the Paris meridian, he was in prison. Years of internal political strife and war postponed the measuring of the meridian, but it was finally done in 1795, and the new metric system was adopted in 1799. The "official meter" was formally interred in a vault near Paris in a ceremony marking the occasion as one "for all people, for all time." Indeed, for the next 150 years, people from all over the world brought their meter sticks to Paris to measure them against the *standard meter bar* and the *standard kilogram block*. Most countries adopted the metric system and the standards invented in France.

In the United States, it has been unexpectedly difficult to legislate the adoption of a metric system. While it would involve unlearning habits of measurement, retooling production machinery, and retraining workers, the long-range advantages are obvious.

Early in your physical science courses, you will be introduced to the system currently in effect, called the SI system (from the French *Système International*). In essence it is still the metric system,

7. The full set is femto-, pico-, nano-, micro-, milli-, centi-, deci-, deca-, hexta-, kilo-, mega-, giga-, and tera-.

but the standards have become much more exact. No longer is the master meter bar the standard for length. By 1983, atomic clocks were so accurate and the measurement of the speed of light so exact that the meter could be redefined, that is, restandardized, as the distance traveled by light in a specific time interval. Time is now defined in terms of a certain number of cycles of radiation in the cesium atom. As will be seen in Chapter 3, masses at the atomic and molecular level are measured in *atomic mass units*, based on the standard mass of a carbon atom (although we still use kilograms as a macroscopic unit of mass). The reason for the change in standards has to do with the tremendous progress in the twentieth century in exploring nature at the atomic level. As physicist Philip Morrison likes to say, "Every atom is a storehouse of natural units, more exact than the International Bureau of Weights and Measures." Except for the standard kilogram, kept in a vault near Paris, it is no longer necessary to refer to one specific standard. Rather, all of the other SI units are based on standards that are replicable wherever there is the equipment and the scientists to do the measuring.

The SI system includes all the units you will be using in science. In physics, in addition, you will have to learn combinations of units, such as newtons (units of force) and joules (units of energy); in biology, base pairs. But, as in all the vocabularies we have been describing in this chapter, no one expects you to master everything at once. These terms appear at a manageable rate, as you move from one course unit to the next.

Units in Astronomy

In astronomy, the term "astronomical unit" is the average distance of the earth from the sun. When Copernicus postulated his theory that the earth orbits the sun, he also calculated the relative distances of the known planets in astronomical units. At that time, no one knew the actual distances of the earth to the sun, or to any planet. Later astronomers were able to measure the distance from the earth to the sun and found it to be 1.4960×10^{11} meters, which we now call one astronomical unit. Another convenient measure used by astronomers is the light-year, which is the distance traveled by light in one year. Once the speed of light was measured, this unit could be calculated, too. It is 9.4605×10^{15}

meters for an average year. The large distances involved in the study of astronomy mean that these special large units have to be used; else the numbers would be just too big.

Achieving Fluency in a New Language

Reading or listening to new scientific terms in lectures alone will not bring about the level of understanding you will get from problem solving (in physics and chemistry) or from a close study of biological processes. Even though there are as many new words in some introductory science courses as in a first-year language course, you will not benefit much (as your friends in foreign language courses may) from merely memorizing new words.

Experts disagree as to whether language precedes thought or thought precedes language, but no one doubts that the mastery of a new vocabulary gives a person access to, and the means of organizing, ever more complex ideas. When your science vocabulary is in place, that is, when you are familiar with the terms, mathematical relationships, and visual images that anchor the concepts you are learning, you will find your mental world expanding. You will think thoughts you never thought before. And when you try to tell someone else what you are thinking, they will understand! As this happens more and more, you can be sure you are well on your way to becoming fluent in a new language—the vocabularies of science.

3

Understanding Science

> When I really understand something, it is as if I had discovered it myself.
>
> *Richard Feynman*

Understanding science involves more than merely learning its vocabulary or knowing certain facts. Especially in the physical sciences, scientific concepts begin to make sense only when students apply them. This is why, particularly in physics and chemistry, your ability to lay out and solve quantitative problems is one route to understanding these subjects. Another important component is laboratory work, where students come closest to doing science; a third is developing your ability to become familiar with, and freely use, the models on which scientific concepts depend.

In a book of this sort, we cannot teach problem solving, any more than we can walk you through the laboratory. But by foreshadowing what lies ahead we can help to guide your approach to understanding science. Chemistry problems are different from physics problems and biology problems don't appear until genetics. But there is something about solving problems (even those that have been solved many times before) that will teach you how scientists think. Every science is experimental; in all laboratory courses, you will have to make observations, take measurements, draw inferences, and write lab notes and reports. So we think the sciences have enough in common to permit us to discuss them all in one chapter.

Most sciences organize their material vertically, so that one concept leads to the next. If you keep up with your assignments, you will always have something to stand on as you try to grasp what is new. Although there are no universal "how-tos" in science, no tricks or reliable shortcuts you can count on to get you through, the good news is this: in laboratory work, and problem solving, as in all other kinds of skills, practice makes hard things easier.

Problem Solving

Early in the fall semester, just after the first hour examination, a student comes to complain to her physics professor about her grade. "How come I did less well than I thought?" the student asks. "Going into the exam, I thought I understood everything quite well." The professor looks over the exam and agrees that she could have done better. "I worked so hard," the student continues, placing the textbook on the professor's desk. "I went over every chapter several times. Look how I've highlighted the text." The professor is impressed. Nearly every other sentence has been marked up in bright yellow. "I memorized every definition, every term," the student concludes. "What did I do wrong?"

The professor realizes that this beginning student doesn't yet know what understanding means in physics. "Did my exam ask you for definitions?" she asks. "Were you asked to explain in general why some object falls with a certain acceleration? Why do you think I put five different problems on the exam? In physics, the only way for both you and me to know that you understand is to see how you tackle problems. That's why we tell students that to study for a physics test they need to solve as many problems—the ones assigned and the ones recommended—as they can."

Then, as she has done many times before, the professor goes to the bookcase and pulls out a volume called *The Science of Swimming*, written by "Doc" Councilman, the famous Indiana

University swimming coach. "I don't think you could learn to swim by reading this book, even if you were to memorize every paragraph." The student laughs. She suddenly understands the point. You have to get into the pool to learn swimming. It's the same in science.

Historical Background

Problem solving is at the heart of understanding the physical sciences and advanced biology, but this wasn't always so. German universities, pioneers in science education 150 years ago, did not include problems either as examples or as end-of-chapter exercises. A glance at any very old textbook in physics or chemistry will illustrate this: except for equations, the entire text is in narrative form.

The British first added problem sets to their science curriculum. First, these were just lists of problems at the end of the book. Later, problems were interspersed throughout. Some historians think that it was the famous "Tripos," the end-of-studies examinations in mathematics at Cambridge University, that made problems part of the teaching of physics and mathematics. Training in physics and mathematics wasn't differentiated in England until the beginning of the twentieth century. Candidates for degrees in both physics and mathematics had to compete in the Tripos by solving artificial problems constructed by a committee of their senior professors, who would sit together for days developing these problems. The professors' problems became known worldwide as real challenges to advanced students. They were named by the year they appeared on the Cambridge University exams: "Tripos 1874," "Tripos 1898," and so on.

It took a long time, as late as World War II, for problems of graduated difficulty to find their way into introductory chemistry and physics texts, but eventually, science instructors came to the conclusion that there was no better way to teach concepts and understanding than through problem sets. Problem solving is still more common in the English-speaking countries. The rest of the world remains under the influence of the older German/French traditions in training scientists (concepts are presented to be learned, not applied). Many science educators believe that it is problem solving that makes U.S. physical science students so competitive in the world at large. They learn to figure things out for themselves.

Approaches to Problem Solving

Since problem solving is considered to be an essential skill, it was only natural that specialists would begin to study the process of problem solving itself. The study of problem solving and of learning theory has, in fact, produced a new subject, the science of *heuristics* (the study of approaches to problem solving). Researchers in this field set up experiments to compare how experts in science solve problems compared to beginners. They want to find out how experts begin to work a problem, what they think about as they work the problem, in what sequence they do certain steps, and what they do when they get "stuck," so they can teach problem-solving skills more effectively to students. It used to be assumed that experts do well in solving science problems because they know more science. The new research challenges that assumption. Experts, the researchers now think, do well because they are experienced problem solvers.

The game of chess is considered an example of continuous problem solving. In observing experts and beginners playing chess, researchers have noticed that good players plan in terms of large blocks of moves—this is called "chunking" in heuristics—and that beginners think only one step at a time.

When these researchers turn to how people solve physics problems in introductory courses, they find that, as in chess, beginners spend less time than experts thinking ahead in whole chunks of steps. Instead, beginners rush to find a formula that looks as if it will fit the problem and then "plug" in numbers. (Students call this "plug-and-chug.") Experts, in contrast, don't do anything very fast. They stand back and contemplate the problem in terms of other problems they have seen before and in terms of fundamental physical principles. Fred Reif, a pioneer in physics heuristics who is at Carnegie Mellon University in Pittsburgh, describes experts' problem solving as involving four sequential steps:

1. Describe the problem as completely as possible, including a drawing.

2. Try to figure out what kind of a problem this is and plan a strategy for solving it.

3. Implement the strategy and finally plug in the numbers.

4. Check the solution for consistency and reasonableness.

If beginning students followed these steps, they would do much better in science.

Problem Solving in Chemistry

Catherine Middlecamp was a chemistry major at Cornell University in the 1970s and holds a Ph.D. in chemistry from the University of Wisconsin. She now directs a chemistry tutoring program at Wisconsin for minority students, who have historically been underrepresented in science. Her experiences in teaching chemistry have produced a self-help guide,[1] and her conviction is that mastery of chemistry is well within every college student's grasp.

Analyzing the Problem

Just as in the four-step process described above, the first thing Middlecamp asks herself when she is faced with a new chemistry problem is: What kind of a problem is this? Have I worked on problems like this one before? If so, what do I remember about this type of problem? In chemistry, she says, there are many problems but only a limited number of types of problems to solve. So mastering problems by type is a good strategy for beginning students. Some chemistry problems test a *qualitative* understanding; for example, the materials present in a compound and their properties. Most problems, however, are designed to test a student's *quantitative* ability; to measure or to calculate the amounts involved in chemical processes. Although the knowledge required to do these quantitative problems is comprehensive and detailed,

1. Elizabeth Keane and Catherine Middlecamp, *The Success Manual for General Chemistry* (New York: Random House, 1986).

the approaches can be classified in terms of the principle of chemistry that is involved.

Like experts playing chess, beginners should spend as much time as necessary "walking around" a problem, describing it and thinking about it, before rushing in to find a formula. In approaching a chemistry problem, you might ask yourself: "Does this particular problem have to do with equilibrium or with a completed reaction?" "Is the reaction gaseous or in solution?" "What are the conditions of the reaction?" Only after thoroughly describing the problem should you begin to think about the answer. A successful student doesn't set up equations or look for formulas before deciding what kind of an answer should probably result. Will it be in kilograms? liters? moles?

Like a pilot trying to land an airplane at an unfamiliar airport, you should make a "pass" at a problem before settling down to solve it by asking:

1. What kind of quantity am I trying to end up with?

2. What will I need to do to get this kind of quantity?

3. Which chemical principles am I going to have to apply?

If you think about what kind of answer you are looking for before starting to work the problem, you are much less likely to go wrong. This is strategic problem solving.

The "Both-Ends" Strategy

Students often fail to take this good advice because, at least at the outset, they find themselves moving too slowly, even going backwards. They are right in a way, because strategic problem solving means that you attack the problem simultaneously at both ends: you continually go back and forth from the question to the kind of answer you're looking for, then back to the question for more clues as to how to get there. A "both-ends" strategy has some psychological benefits. Instead of a finding process—having to find an elusive answer in a sea of possible answers—it involves a fitting process—fitting the answer to the question. Professional scientists tell us this is the method they use in solving real-world problems in science.

The Assessment Strategy

Another approach is to assess the usefulness of the information given in the problem. Some problems are overdefined, that is, there is more information given than is needed. Here the challenge is to figure out what is important and what is not. Some problems appear to be underdefined, that is, too little information is given. Most problems are a mix of both. Very often the student has to do some digging before setting out to find the solution. You might not be told the name of the solvent involved and thus think the problem is underdefined, until you realize that you don't need to know the kind of solvent in order to solve the problem in the first place.

Deciding on your strategy is half the battle. First, analyze your own thinking patterns. If "getting started" is the hard part, digest the examples in the text and in lecture notes thoroughly before starting on the problem set. If "getting stuck" is a problem, return to the problem hours later, if necessary, for a retry. Meanwhile do as many unassigned problems as possible for extra practice. (This is good advice for physics problems, too.)

Special Difficulties

Two factors create special difficulties in chemistry. First, the microscopic world cannot be experienced directly. Chemists therefore have to work indirectly—by means of reactions, purification, and quantification—to get a sense of what is going on at the molecular levels. We are beginning to have techniques capable of measuring a "molecular event" (see Chapter 5). But in your labs, what is happening at the atomic and molecular levels in chemistry and biochemistry can still only be inferred. Second, chemistry requires a precise attention to quantities. Chemistry can be like bookkeeping, says Middlecamp: "Sometimes you bookkeep electrons; sometimes you bookkeep bonds."

More than in your other introductory science courses, chemistry problems will be connected to your laboratory work. What happens when the chemist adds manganese oxide to iron is a question that is answered both theoretically and in the laboratory. That's why the introductory laboratory is a very important component of chemistry and an integral part of your course.

Professional chemists are very committed to problem solving, both in their own work and in training others. Roald Hoffmann, whom we interviewed for this book, is a Nobel laureate in chemistry who studies the shapes and motions of molecules. He won his Nobel Prize in 1980 for work on *frontier orbitals,* the chemists' name for the outermost shell available to electrons surrounding the core of a molecule. When we asked him what role problem solving plays in understanding science, he made an analogy to learning a foreign language. He said:

> You won't really know if you understand the vocabulary, grammar, and idioms of a foreign language, if you only read it or do vocabulary exercises. . . . It is only when you have to translate an entire sentence or paragraph from your own native language into the one you are studying that your knowledge is really tested. . . . That's the same test you get in solving problems in physical science. A well-constructed problem set should challenge you to bring together everything you know.

Problem Solving in Physics

Teachers of physics disagree on many things: Which demonstrations should be included in a course in introductory physics? Which topics should be left out or emphasized more? How much and what kind of laboratory work should be included? How many quizzes? What kind of exams? But one thing on which there is no disagreement is that understanding physics means being able to master increasingly difficult problems.

So important has problem solving become in the teaching of physics at college that physics instructors select (or reject) new textbooks not so much for their narrative, illustrations, or reputation, as for the quality, level, and variety of their end-of-chapter problems. That's because, as college physics professors tell us, understanding physics demands the ability to analyze a physical situation that one has not encountered before in terms of its fundamental attributes and to solve the problem quantitatively.

In chemistry, the problems in textbooks apply to a particular material (molecules or compounds). That's because chemistry is the study of the properties of specific substances and reactions. The laws of physics, in contrast, apply universally. Whether a force is acting on an animal, a ferris wheel, or a block on an inclined plane, the situation will be analyzed in terms of general laws of mechanics: the masses, their initial states of motion, and the forces acting on them. In an early chapter of a widely used textbook in introductory physics, there is a problem in which the student is asked to find the *center of mass* of an ammonia molecule. The very next problem asks the student to find the center of mass of the earth-moon system. And the one after that concerns the center of mass of a homogeneous semicircular plate. The three problems involve chemistry, astronomy, and carpentry, respectively. Yet they all have to do with the same concept (center of mass) in physics. This is what makes physics unique, interesting, and very powerful.

Simplifying Problems

The first rule of solving physics problems is to try to turn a complex problem into a simpler one. Here's an example from kinematics—the study of motion without regard to the cause of motion—that appears early in a first course in physics.

> A ball is kicked at a speed of 24 meters per second at an upward angle of 45 degrees from a level playing field. A player on the other team is 75 meters away. How fast should he or she run (i.e., at what constant speed) to catch the ball at about arms' height, or 1.8 meters off the ground? Neglect air friction.[2]

An experienced physics problem solver would not begin at the beginning of the problem but at the end: the speed of the receiver. How is speed usually calculated? Distance divided by time. What is the distance in this case? What is the time? What information will help me find the distance? What information will help me find the time? At this point, the problem has become an ordinary two-step problem: finding the horizontal distance to the point where the ball descends to a height of 1.8 meters, and the time of flight to that point.

One of the obstacles facing students of physics is learning to

2. R. Resnick, D. Halliday, K. Krane, *Physics*, 4th Edition (New York, John Wiley, 1992), p. 72.

solve problems that seem new but that are actually, in important respects, similar to ones they have already learned to solve. But how does one recognize similar problems in physics? The best way is to learn to think algebraically instead of arithmetically. Take the example of the ball being kicked upward from the playing field. You could think about the speed at which the ball takes off as 24 meters per second as given in the problem. Or you could resist the urge to plug and chug, and think, rather, about the ball's speed as its initial velocity, written symbolically (you will learn this notation) as v_0. Likewise, you could think about the ball's projection angle of flight as 45 degrees from the ground, as given in the problem, or, more abstractly, that is, algebraically, as making an angle, θ (*theta*) with the ground. In the same manner, the height at which the ball is to be caught, 1.8 meters, could be written symbolically as h.

Writing the numbers down early in the process of solving the problem tempts you into treating the problem as a simple calculation. But if you translate the physical conditions into algebraic symbols, you will begin to notice similarities between this specific problem and more general problems of *projectile motion*. You will find yourself thinking about physics problems the way physicists do instead of turning it into a problem in applied mathematics. Students who think physics has to do only with numbers miss out on many of the insights and much of the grandeur that physics can provide.

The algebraic approach has another advantage in problem solving in physics; it is easier to catch an algebraic error than one in numerical calculation. If, for instance, in calculating a time of flight in the above problem (or in any problem involving speed), you arrived at the expression $t = v/d$ (time = velocity divided by distance), you would know right away that you had reversed v and d and needed to backtrack. But if you had been working with the numbers from the beginning, your answer might seem reasonable even if your reasoning was wrong.

How can you tell whether your reasoning is correct and whether the form of the expression makes sense? It helps to apply some "what-ifs" to problems in physics. If the force of *friction* increases, what should happen to the acceleration? Sometimes, a physicist will take one or more of the *parameters* (variables) of a problem to an extreme—for example, allow the mass to approach infinity—just to see what happens to the quantities involved.

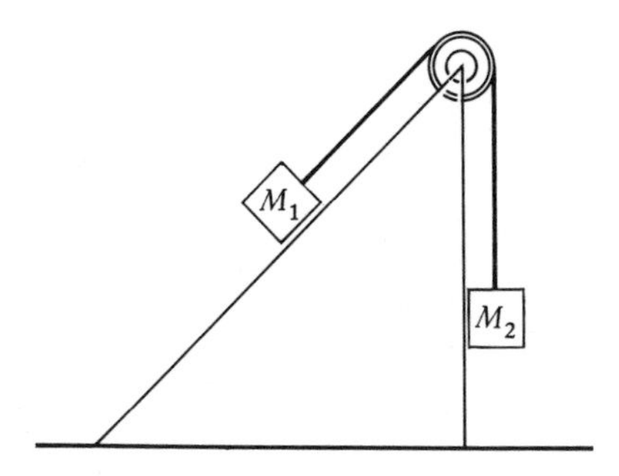

FIGURE 3.1

Sketch of actual physical environment. One block (M_1), is moving upward on an inclined plane. The other (M_2), is attached to a string over a pulley.

Physical phenomena take place in a three-dimensional world. Yet diagrams in two dimensions help students see what's going on in the physical world. Therefore, no discussion of problem-solving in physics can be complete without some discussion of diagrams.

The Free-Body Diagram in Mechanics

Another skill you will need to master in solving mechanics problems is the construction and use of the free-body diagram. The free-body diagram is rooted in a very basic strategy that physicists use to analyze the complex world of motion. If you understand the purpose of the free-body diagram, you will begin to appreciate its power. Its purpose is to focus your attention on what physicists call *local description*.

Let's see how a free-body diagram clarifies a typical physics problem in introductory mechanics; how to determine the acceleration, that is, the change in velocity of two blocks tied together by a light string over a light pulley (Figure 3.1).

In constructing the free-body diagram, you are to diagram all the forces acting on an object. The first time around, you are to ignore the mass of the pulley, so there are actually two moving objects in this system. You need to treat these two objects separately, drawing a free-body diagram for each. The diagram of forces, if done correctly, recreates exactly the environment for an

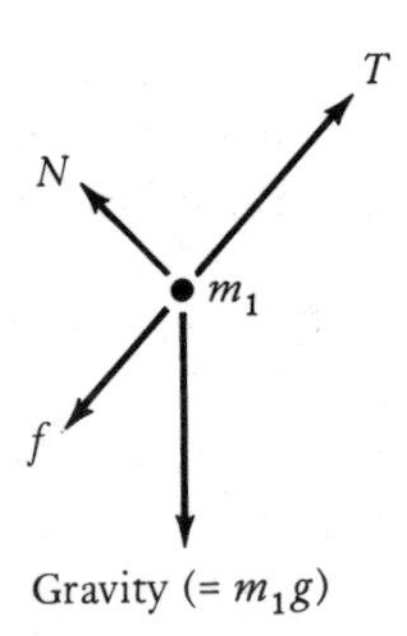

FIGURE 3.2
Free-body diagram of the block (M_1).

object as if it were in free space. This is what the term "free-body" means, removed from its real-world surroundings, including the earth itself. When you apply this idea to the block moving up on the inclined plane (M_1), its free-body diagram is shown in Figure 3.2. A free-body diagram of the hanging block (M_2) is shown in Figure 3.3.

The construction of a free-body diagram replaces the actual, physical environment with a collection of forces. The reason the analysis is done in terms of forces is central to Newtonian mechanics. After Newton, force became the language of motion. $F = ma$ dictates that all *causes* of any change in velocity will be expressed in terms of F. In this particular problem, there are four forces at work. The force of gravity is one; friction (labeled f) is

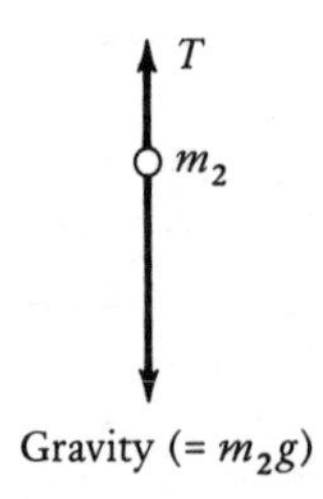

FIGURE 3.3
Free-body diagram of the block (M_2).

a second. These are obvious. Less obvious, but integrally related to Newton's way of describing motion are two additional forces: the perpendicular force (usually called the normal force (*N*)), the force exerted on the block by the inclined plane at right angles to the plane; and the tension force (*T*), exerted by the string.

You might not think of *N* and *T* as forces, but rather as constraints on the block. After all, the block cannot penetrate the inclined plane; nor can it "run away," because it is being held by the string. However, to the physicist, constraints have to be described in terms of forces. So two additional forces replace these constraints: the force on the block by the inclined plane, and the force on the block by the string. You'll notice that on the free-body diagrams in Figures 3.2 and 3.3, each of the forces is written as an arrow, indicating its direction.

You can begin to see how powerful a tool the free-body diagram is. As the diagram makes very clear, whatever the block is in contact with in its immediate vicinity determines its change in motion. This is what is meant by local description.

Action at a Distance

The force of gravity is not in the immediate vicinity of the block. You may have been told somewhere along the way that gravity, unlike other forces, operates at a distance from an object. The current view (held since the middle of the nineteenth century) is, rather, that gravity is best imagined as a *field* surrounding the earth. One way to imagine this gravitational field is to think of an infinite number of "arrows" of force, one arrow per point in space around the earth. To measure gravity at a particular point in space (due to different attracting objects, such as the earth, moon, and sun), the physicist "adds up," so to speak, the total number and direction of those arrows at that point.

This way of conceptualizing action at a distance, such as electric and gravitational forces (both of which appear to act over distance), was the brainchild of Michael Faraday (1791-1867), a blacksmith's son who had very little formal education in mathematics. Despite his modest beginnings, Faraday's research at the Royal Institution in England led, eventually, to the theory of electric and magnetic fields, a theory completed in 1873 by physicist James Clerk Maxwell. At the time Faraday was beginning to work on electric attraction and repulsion, scientists already under-

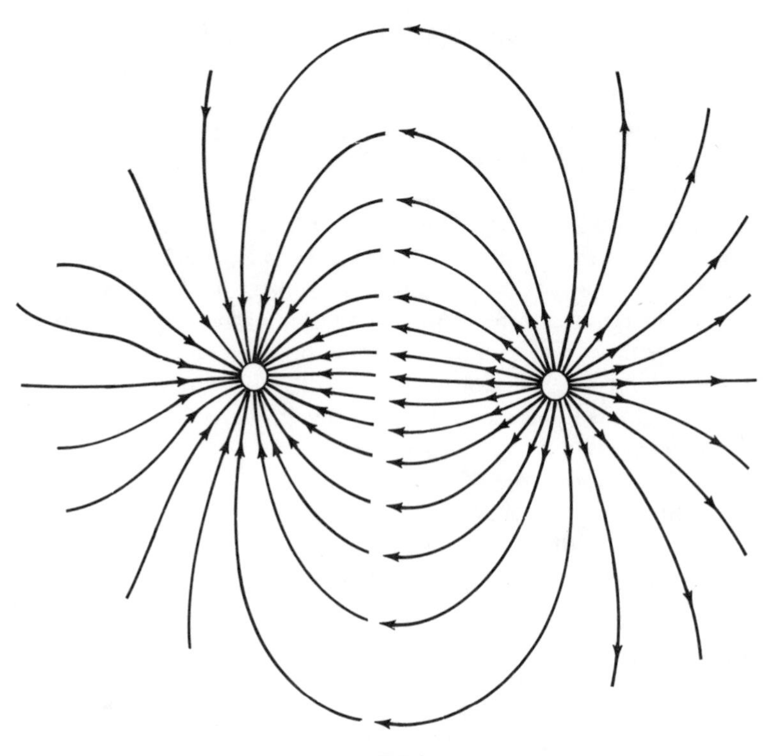

FIGURE 3.4
Model of electric force at work.

stood that those relations could be expressed mathematically as Coulomb's law:

$$F = k \frac{q_1 q_2}{R^2}$$

Where F is the force of interaction (attraction or repulsion), k is a constant, q_1 and q_2 the quantities of electric charges, and R the distance between the charges.

Faraday wanted something more than an equation. His own mathematical background was poor, and he needed something concrete to imagine. So he constructed in his mind another model, one that would allow him to "see" electric force at work. He imagined lines (tubes) of electric force filling the empty space between interacting bodies (Figure 3.4). He imagined the same

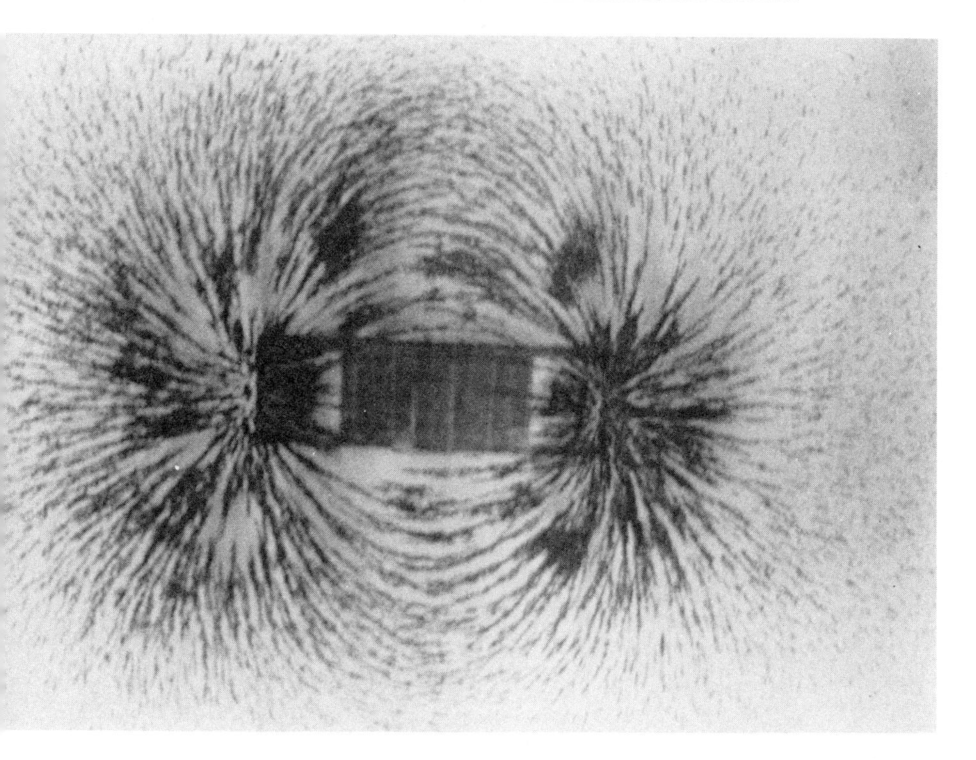

FIGURE 3.5
Photograph of iron filings around a bar magnet.

for forces of a magnet similar to the patterns made by iron filings when they distribute themselves around a bar magnet (Figure 3.5).

Action-at-a-distance force was understood as the push and pull of *field lines*. Today, the idea of the field (although it has become considerably more abstract than in Faraday's time), pervades all of physics. Gravity itself is understood to be a field and therefore is no longer considered action at a distance but rather a kind of contact force.

Once you understand that gravity can be described locally, you will be well advised, in solving introductory physics problems, not to go beyond local description. If you become distracted by forces not in the immediate vicinity of the problem, you will get into trouble. Suppose, in Figure 3.1, you started wondering about the origin of the tension on M_1. You might confuse the weight of M_2 with the tension force of the string. But not all of the weight of M_2 is producing acceleration on M_1; some of its own weight is acting on M_2 itself. Your best strategy is to ignore everything except what is happening locally. This will keep you from making wrong force assignments.

Approaching Chemistry

A famous physicist, Richard Feynman (1920-1989), who is quoted at the beginning of this chapter, once posed an interesting speculation: Suppose the present civilization, including all known science, were about to be destroyed, and we could leave only one piece of information from which future generations could entirely reconstruct our store of knowledge. What should that piece of information be? Feynman's choice was the statement, "All matter is made up of atoms." Why is that piece of information so vital to our understanding of nature? Feynman would say because this one idea cost our forebears their greatest effort and took humankind the longest time to construct and to confirm. Modern chemistry is firmly based on this atomic model.

For an understanding of chemistry, it is vital to grasp three key concepts: (1) the *mole,* a standard unit for measuring large quantities of molecules and atoms; (2) *atomic mass,* a quantity assigned to every element; and (3) the underlying logic and pattern of the *periodic table.* All three are related, both to one another and to the fact that matter is made up of atoms.

Chemistry of Elements

Until the end of the Middle Ages, questions about nature were considered to have been settled by the great systematic work of the Greek thinker Aristotle (384-322 B.C.). In Aristotle's system, the entire universe was made up of five "elements," (not the same as the chemical elements known today). Four of them—earth, water, air, and fire—are concepts that are familiar to us; the fifth—quintessence—made up the heavenly bodies (planets, stars, and the sun). Each of these elements had a place in the universe and a specific "tendency": the tendency of a rock, which is mostly earth, was to fall to the bottom of a pond; that of air bubbles was to float above water; that of fire was to rise above the air; and that of the heavenly bodies was to travel endlessly in their circular orbits. "Natural motion"—distinguished from "violent motion"—didn't need to be accounted for. Sustained force was needed only to displace some element (violently) from its natural position. In this worldview, as you can see, matter and motion—what we presently call chemistry and physics—were not yet differentiated.

Modern chemistry has identified 92 elements in nature, and over a dozen that have been artificially created at the heaviest end of the periodic table. We consider these elements to be the building blocks of molecules and compounds, which in turn are the building blocks of all the objects we know—probably everything in the universe as well. You do not need a course in chemistry to know that a water molecule is made up of two hydrogen atoms and one oxygen atom, represented symbolically as H_2O. But even this basic information had to await a revolution in chemistry, one that continued well into the middle of the nineteenth century. That revolution had its beginning in the experimental work of Joseph Gay-Lussac (1778-1850), a French chemist, and the theoretical speculation by Amadeo Avogadro (1776-1856), an Italian physicist. You won't get through introductory chemistry without learning Gay-Lussac's law of gases or memorizing Avogadro's number, the fixed number of molecules in certain volumes or weights of elements.[3] But you may not be taught enough of the history of chemistry to appreciate how atoms and molecules came to be understood.

By Gay-Lussac's time, chemists had learned to isolate some elemental gases, to measure them by volume in isolation, to get them to react, and then to measure the resulting products. For example, when Gay-Lussac experimented with oxygen and hydrogen, measuring these gases at the same temperature and volume, he found that two volume units of hydrogen and one volume unit of oxygen produced two (not three) volume units of water vapor, as expressed schematically in Figure 3.6.

This was not expected because, according to the earlier atomic theory of Dalton, the most elementary form of hydrogen gas was

3. That certain volume is 22.4 liters at 0° Celsius and one atmosphere; and that certain weight is equal in grams to the molecular weight.

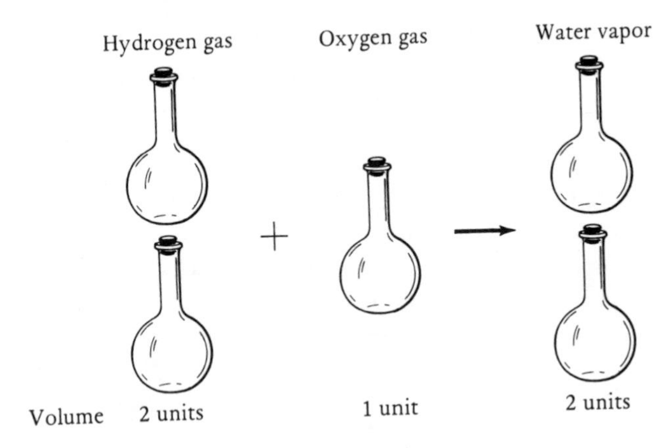

FIGURE 3.6

Schematically expressed results of Gay-Lussac's experiment with oxygen and hydrogen.

a single hydrogen atom, and the most elementary form of oxygen gas was a single oxygen atom. If this were the case, the relationship at the atomic level would be as expressed in Figure 3.7.

How could two units of water vapor be made up of an entity of hydrogen and a half-entity of oxygen? Chemists of the day believed that atoms are indivisible. Avogadro suggested that per-

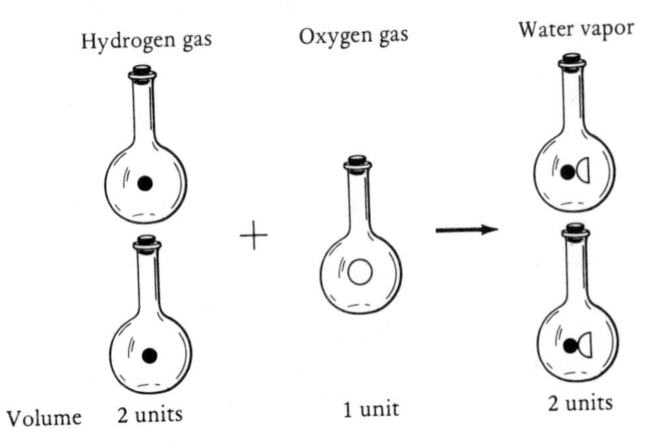

FIGURE 3.7

The symbols within the flasks represent individual atoms.

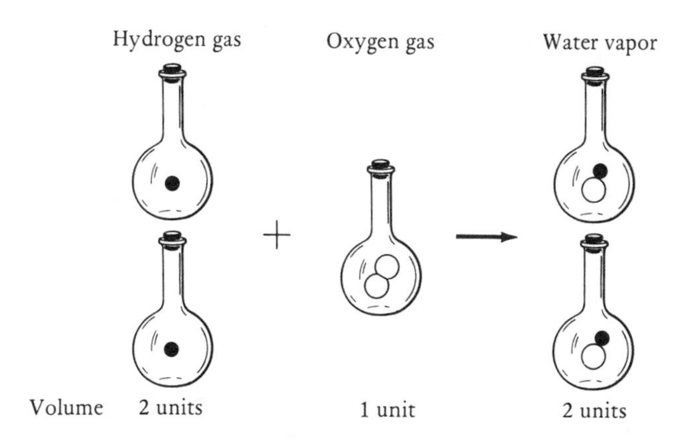

FIGURE 3.8
The symbol within the oxygen gas flasks represents a molecule made up of two oxygen atoms.

haps some elements do not exist in nature as single atoms, but rather as groups of atoms; without knowing it, he was describing the molecule. If elementary oxygen exists as a molecule made up of two oxygen atoms (expressed as O_2), Figure 3.8 represents the oxygen-hydrogen reaction, and the end product is two units of HO.

However, similar experiments showed that hydrogen is also a two-atom molecule (represented as H_2). Therefore, what is actually happening at the molecular level when oxygen and hydrogen gas combine is shown schematically in Figure 3.9 and is expressed by the equation:

$$2H_2 + O_2 \longrightarrow 2H_2O$$

The integers in front of the chemical symbols (as in algebra the number 1 is omitted) stand for the number of volume units. So $2H_2$ means two volume units of two-atom hydrogen molecules and O_2 means one volume unit of two-atom oxygen molecules. Avogadro's insight was that in a chemical reaction the only important quantity is the number of molecules, not the volume of either the reactants or the resultant product. We need a concept called "mole" to put this together.

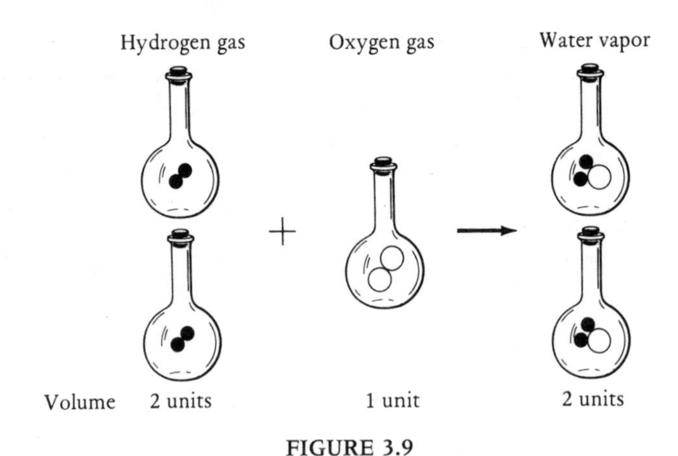

FIGURE 3.9
Both hydrogen and oxygen are two-atom molecules.

The Mole

Like the chemical formulas for single substances (H_2O, CH_3OH), chemical reaction equations have both a *microscopic* and a *macroscopic* (visible to the naked eye) meaning. All chemical reactions take place at the molecular (microscopic) level; atoms and molecules are exchanging electrons; positive ions are attaching to negative ions. But, as we have noted, chemists cannot physically handle substances on the atomic scale; they are too small to weigh and too numerous to count. A single speck of dust has 10^{16} atoms. (That's 1 followed by 16 zeroes.) With entities this small, there is no way to work in a chemistry laboratory with individual atoms. Yet it's at the atomic and molecular levels that reactions take place. The problem is: how do we get from atoms and molecules to measurable real-world quantities?

What chemists needed was a bridge to connect microscopic entities such as atoms or molecules with measurable quantities such as ounces or grams. Experimentation, enhanced by imagination and insight, provided that bridge. It's called the mole, and although your chemistry text may try to make the term more "user-friendly" by calling it a chemist's dozen, do not be misled; it is one of the most important ideas in chemistry.

The definition of a mole is very orderly and precise. One mole of any chemical element or compound is the amount of that substance containing as many atoms or molecules as there are atoms in 12 grams of pure carbon 12.[4] The number of atoms in 12 grams of carbon 12 is extremely large; so large, in fact, that one mole of any element or compound is most conveniently written in scientific notation, as 6.02×10^{23}. This means that when you are asked to take one mole of atoms or molecules, either in laboratory work or in solving problems in chemistry, it means 6.02×10^{23} atoms or molecules, called Avogadro's number, in recognition of Avogadro's insight.

Atomic Mass

What you see listed in your chemistry textbook as *atomic mass* or *molecular mass* (sometimes called *atomic weight* or *molecular weight*), means the mass in grams of one mole of atoms or molecules. Large masses of elements or compounds are expressed in moles; small masses in atomic mass units, written as u. If there is one mole (Avogadro's number) of atoms in 12 grams of carbon 12, then each of these carbon 12 atoms has a mass of 12 u. If that sounds confusing, think about it this way: Avogadro's number is simply a way of converting atomic mass units to grams.

Without modern instruments, nineteenth-century chemists estimated the atomic masses of the then-known elements by very carefully measuring the weights of reacting substances, before and after a reaction. Since they had to rely only on ratios of reacting masses, any mistake in the formula for the reacting substance would result in incorrect atomic masses. For a long time, before Avogadro's insight was generally accepted, chemists confused molecular with atomic masses, and so the determination of accurate atomic masses was very long in coming.

Today, the calculation of atomic masses is no longer based on conjecture or experimental trial and error, but on mass spectrometry, a measurement technique developed in the first quarter of

4. Carbon 12 is an isotope (a slight variant) of standard carbon. The 12 in carbon 12 is the number of nucleons (protons and neutrons) in the nucleus of carbon 12. The reason 12 grams of carbon 12 is chosen to be the standard for the atomic mass unit is because carbon is the most abundant solid element on earth and, because of this isotopic abundance, carbon 12 is easiest to obtain in isotopically pure form.

$$\text{methane} + \text{oxygen} \longrightarrow \text{carbon dioxide} + \text{water}$$

$$CH_4 + 2\,O_2 \longrightarrow CO_2 + 2\,H_2O$$

$$1 \text{ molecule} + 2 \text{ molecules} \longrightarrow 1 \text{ molecule} + 2 \text{ molecules}$$

$$1 \text{ mole} + 2 \text{ moles} \longrightarrow 1 \text{ mole} + 2 \text{ moles}$$

$$\underset{(12+4)}{16 \text{ grams}} + \underset{2 \times (16 \times 2)}{64 \text{ grams}} \longrightarrow \underset{12 + (16 \times 2)}{44 \text{ grams}} + \underset{2 \times (2 \times 1 + 16)}{36 \text{ grams}}$$

$$(\text{sum}) \;\; 80 \text{ grams} = 80 \text{ grams}$$

FIGURE 3.10

Example of a complete chemical reaction, in molecules, moles, and grams.

this century, and refined ever since. But one thing hasn't changed: the principle that the connection between atomic mass units at the microscopic level and grams (and kilograms) at the macroscopic level is expressed in terms of moles.

You might be asked to do the following theoretical calculation: take one mole of methane (CH_4) and mix it with two moles of oxygen (O_2). The periodic table will give you the atomic mass units, which, when converted into gram equivalents, result in 16 grams of methane and 64 grams of oxygen. The reaction is illustrated in Figure 3.10. This is how chemical reactions are supposed to be understood: at the gram level, at the molecular level, and in terms of moles.

The Periodic Table

Through much of the nineteenth century, experimental chemists were busy discovering and identifying more and more elements. As these elements came to be known, chemists suspected that elements bore some quantitative relation to one another, and that groups of elements had similar chemical and/or physical properties. Once chemists were able to calculate atomic masses more accurately, they began to think that all the known elements might be *sequenced* in order of their atomic masses in some chart or table. They were almost right, but it wasn't until 1913, when a young British physicist, H.G.J. Moseley, discovered that certain X-ray lines characteristic of each element were displaced according to their atomic numbers, that the sequencing by atomic number was experimentally demonstrated.[5] Not until the discovery of protons and neutrons, did scientists understand the meaning of the sequencing itself.

Today, the sequencing system is well understood and is routinely displayed in the chart we call the periodic table of elements (Figure 3.11). You won't have to go far in chemistry to locate a periodic table. It will face you in your lecture hall; it will be reproduced on inside covers of your textbooks; you can buy laminated pocket-sized versions of it to carry around in your wallet. It is, no doubt, buried in every time capsule, in the spirit of Professor Feynman's "legacy to posterity."

But its construction took decades to achieve. Imagine a jigsaw puzzle with 30 out of 90 pieces missing and no clue as to whether it is to be put together on the basis of one or several principles of organization. Chemists began with what they had to work with, empirical laboratory results. Based on these results, they tried to group elements by the reaction patterns of certain elements, that is, what kinds of compounds they formed with other elements; their densities; their boiling points; their melting points; and other chemical and physical properties. It was difficult and painstaking, but it was also exciting. One textbook describes that process of sorting out the relations among elements as follows:

> Lithium and sodium . . . have very similar properties. Fluorine and chlorine are also very similar to each other, but they are very different from lithium and sodium. . . . When the first 17 known elements were arranged in the order of their atomic masses, six elements were found to fall between lithium and sodium. What an exciting moment it must have been when

5. The atomic number, printed immediately above the element's symbol in the periodic table (Figure 3.11), represents both the number of protons in the nucleus, and the number of electrons in the outer shell of the neutral atom. The position and number of an atom's electrons, particularly those in its outermost shell, determine many of its chemical properties, its reactivity, and its bonding preferences. But nineteenth-century chemists did not know about protons or electrons, so we have to marvel that they could and did develop any kind of sequencing of elements without this knowledge.

Period	I.	II	III	IV	V	VI	VII.	VIII		VIII	I	II	III	IV	V	VI	VII	Noble gases
1	1 H 1.008																	2 He 4.003
2	3 Li 6.939	4 Be 9.012											5 B 10.811	6 C 12.011	7 N 14.007	8 O 15.999	9 F 18.998	10 Ne 20.183
3	11 Na 22.990	12 Mg 24.312											13 Al 26.982	14 Si 28.086	15 P 30.974	16 S 32.064	17 Cl 35.453	18 Ar 39.948
4	19 K 39.102	20 Ca 40.08	21 Sc 44.956	22 Ti 47.90	23 V 50.942	24 Cr 51.996	25 Mn 54.938	26 Fe 55.847	27 Co 58.933	28 Ni 58.71	29 Cu 63.54	30 Zn 65.37	31 Ga 69.72	32 Ge 72.59	33 As 74.922	34 Se 78.96	35 Br 79.909	36 Kr 83.80
5	37 Rb 85.47	38 Sr 87.62	39 Y 88.905	40 Zr 91.22	41 Nb 92.906	42 Mo 95.94	43 Tc (99)	44 Ru 101.07	45 Rh 102.91	46 Pd 106.4	47 Ag 107.87	48 Cd 112.40	49 In 114.82	50 Sn 118.69	51 Sb 121.75	52 Te 127.60	53 I 126.90	54 Xe 131.30
6	55 Cs 132.91	56 Ba 137.34	57 La 138.91	72 Hf 178.49	73 Ta 180.95	74 W 183.85	75 Re 186.2	76 Os 190.2	77 Ir 192.2	78 Pt 195.09	79 Au 196.97	80 Hg 200.59	81 Tl 204.37	82 Pb 207.19	83 Bi 208.98	84 Po (210)	85 At (210)	86 Rn (222)
7	87 Fr (223)	88 Ra (226)	89 Ac (227)	104 Rf (?) (259)	105 Ha(?) (260)													
	58 Ce 140.12	59 Pr 140.91	60 Nd 144.24	61 Pm (145)	62 Sm 150.35	63 Eu 151.96	64 Gd 157.25	65 Tb 158.92	66 Dy 162.50	67 Ho 164.93	68 Er 167.26	69 Tm 168.93	70 Yb 173.04	71 Lu 174.97				
	90 Th 232.04	91 Pa (231)	92 U 238.03	93 Np (237)	94 Pu (242)	95 Am (243)	96 Cm (247)	97 Bk (249)	98 Cf (251)	99 Es (254)	100 Fm (253)	101 Md (256)	102 No (253)	103 Lr (257)				

FIGURE 3.11
Periodic Table

> it was recognized for the first time that six elements also separated fluorine and chlorine, and several other pairs of similar elements.[6]

Repeating patterns of elemental properties suggested periods and periodicity. Thus, the chart is called the periodic table.

The German chemist, Johann Wolfgang Döbereiner, thought he had found a pattern of triads, groups of three elements that form similar compounds and in which the atomic mass of one is about the average of the atomic masses of the other two. The English chemist, J.A.R. Newlands, proposed that if the elements are arranged in order of increasing atomic mass, the eighth is like the first, the ninth is like the second, and so on. He called this rule the "laws of octaves," but, although his observations were very promising, he was initially ridiculed for his idea. This was about the situation when the great nineteenth-century Russian chemist, Dimitri Ivanovich Mendeleev (1834-1907), began to construct the first modern periodic table.

Only about 65 of the 92 naturally occurring elements were known in Mendeleev's time, and there was some dispute about whether 5 of them were actually elements. Still, on the basis of what was known about the elements, Mendeleev began to arrange the elements in vertical columns based on their similarity of chemical properties, and in horizontal rows in order of their *combining weights* (that is, their atomic masses), so that their periodicity was apparent. Where there appeared to be "missing" elements, that is, gaps in the gradations of atomic masses or in certain of the groups, Mendeleev was so confident in his sequencing system that he actually left specific boxes empty, assuming (correctly) that in time they would all be filled.

For instance, in Mendeleev's table, there are two empty boxes between zinc (Zn, atomic mass of 65.4), and arsenic (As, atomic mass of 74.9). For the second of these two empty boxes, Mendeleev predicted not only the existence of the element germanium (Ge), but much detail about it: its atomic mass, the chemical form of its oxides, the properties of its organometallic compounds, their boiling points, and their densities—all this detail

6. T. Moeller, J.C. Bailar, Jr., J. Kleinberg, C.O. Guss, M.E. Castellion, and C. Metz, *Chemistry with Inorganic Qualitative Analysis*, 3rd. Edition (New York, Harcourt Brace, 1989), p. 160.

about an element that chemists had not yet discovered, much less subjected to laboratory analysis or experiment. Six years later, germanium was discovered with an atomic mass of 72.6, and, in time, the other element, gallium (Ga) with an atomic mass of 69.72, was also identified. Both elements play an extremely important role in modern electronics.

The periodic table is still called the Mendeleev table in Russia, even though the current chart differs from Mendeleev's in a few significant ways. First, the empty boxes are now completely filled. The only "new" elements will be created in the laboratory. The reason that these elements no longer occur in nature is that their *half lives* of radioactive decay are much shorter than the age of the earth.[7] As in Mendeleev's day, the elements are not all arranged strictly in increasing order of atomic mass. There are exceptions, such as argon (Ar) and potassium (K). But the periodic table is essentially the way Mendeleev left it. What is new is that we now know (as Mendeleev could not), why certain elements are similar to other elements; why certain elements combine; and yet why every element is unique.

The Quantum Mechanical Model

About two decades after Mendeleev died, physicists developed the *quantum mechanical model* which is the basis for both the periodic table and for modern chemistry. We know now that the atom is not the smallest particle of matter. The atom itself has a structure; a nucleus—made up of neutrons and positively charged protons—and a number (equivalent to the number of protons) of negatively charged electrons, surrounding the nucleus in a kind of cloud. The number of protons or the corresponding number of electrons determines the atomic number of an element and its position in the periodic table; the sum of the number of protons and neutrons, its approximate atomic mass. Where the electrons are "located"—in which "orbits" or "shells" in the cloud—is determined by the total number of electrons in a particular atom and by the "rules" of quantum mechanics. This is all very important to chemistry since we now know that chemical reactions take place by the

7. Half life is the scientists' way of conveying the death-rate of a radioactive element; specifically the time it takes for one-half of the total number of atoms to decay. This doesn't mean that the other half will disappear in the same length of time; rather, half of the remaining half will do so, and so on.

rearrangement of electrons in atoms' outermost shells. In fact, the number of electrons in an atom's outermost shell and how completely occupied the shell is generally determine its chemical properties.

What determines the number of shells and the number of electrons in each shell has to do with (1) the total number of electrons (and protons) in an atom; (2) the rules governing the number of electrons per atomic shell; and (3) the Pauli principle which states that no two electrons can occupy the same state in any one atom. An electron's *state* is characterized by four conditions called the *quantum numbers*: its total energy, its angular momentum, its magnetic moment, and its spin, according to the quantum theory.[8]

Once you grasp at least the general principles of quantum mechanics and some of the rules governing electron states, you will understand why the quantum mechanical model explains chemical bonding in ways that were not possible before. The noble gases, for example (so-called because they are chemically inert) have filled-up outermost shells. Thus, they have no extra electrons to give or extra spaces to fill, and therefore don't combine. The elements lithium and sodium, though distant from each other on the atomic number sequence (atomic numbers 3 and 11, respectively), share certain chemical properties since in both cases there is one extra electron moving outside of a filled electron shell. The very "periodicity" that fascinated nineteenth-century chemists—the recurrence of certain chemical characteristics along the periodic table—is entirely explained by electron states.

Using the Periodic Table

Why is the periodic table so important? If you had a large number of elements, all with uniquely different chemical properties and no similarities among them, it would be next to impossible to make any predictions about the results of untried chemical reactions or of the properties of newly synthesized compounds. It is the regularity and similarity among elements in the same vertical column of the periodic table, such as the alkali metals Li (lithium), Na (sodium), K (potassium), and Rb (rubidium) in Group I, and

8. Chemists call the four characteristics: the principal quantum number, the subshell quantum number, the orbital quantum number, and the spin.

the halogens F (fluorine), Cl (chlorine), Br (bromine), and I (iodine) in Group VII that make chemistry more systematic and predictable.

Once you become more familiar with it, the periodic table provides clues as to the ratios of combination. For example, Na (sodium) from Group I and Cl (chlorine) from Group VII combine to form sodium chloride (NaCl)—table salt—in a one-atom-to-one-atom ratio. From this we can predict that other salts, such as KCl, LiI, and so on, combine in the same one-atom-to-one-atom proportion as NaCl. Naturally, there are some variations within each group as to quantitative properties, such as the ionization energy (the energy required to strip an atom of its outermost electron). Here's where the column and row arrangements are particularly valuable. As you go from top to bottom of a column within the same group, ionization energy decreases; as you go from left to right across periods, ionization energy increases. These and many other properties depend on (and can be learned from) the element's position on the periodic table, making chemistry a more predictable science.

More than anything else, the periodic table is the symbol of chemistry. Years of training help engrave the periodic table into the chemist's memory. As a student, you will be encouraged to read the table, but you will never be expected to memorize it. It will always be available for reference when you do your homework and when you take exams.

The Chemistry Laboratory

To a greater extent than is true in introductory physics, lab work is vital to understanding chemistry and, therefore, is integrally related to the introductory chemistry course. In fact, your chemistry laboratory assignments are going to be so closely linked to your lectures and textbook that you will find that you cannot master either without the other. Your study of solution chemistry and equilibrium, for example, will not make much sense to you until you do the experiments assigned in chem lab. Conversely, the hands-on applications of chemistry in the lab will give you the insights you cannot get from textbook assignments alone or from your lecture notes.

Most of your lab work in introductory chemistry will have you doing partly qualitative, partly quantitative analysis. Much of the time you will be mixing *inorganic* (noncarbon) substances; sometimes, though, your reagents (chemicals) will be *organic* (containing hydrocarbons). The introductory laboratory will give you a hint of what it would be like to be an experimental chemist—the joy that comes from the unexpected reactions that may occur. Even when the lab work is more of an exercise, it will be exciting to discover for yourself that there is a reality to the symbolic shorthand and the theoretical explanations given in class.

Even at the beginning levels, laboratory chemistry gives you a sense of the discipline of mind and hand that a working chemist develops. You will be asked to prepare a chart of various *anions* (negative ions such as Cl^-) and *cations* (positive ions such as K^+), placing them in groups according to the similarity of their properties. The goal is to find ways of screening out certain ions and detecting others. From laboratory work, you will learn to manipulate chemical equilibria and to control precipitation. You will also learn not to do anything without observing safety guidelines. You will gain a sense of power in seeing the results of what you do.

At many universities around the country, professors of chemistry employ their undergraduate students in their own professional laboratories to assist them in their ongoing research. Students may test materials that might be recycled, synthesize new compounds, be involved in pharmacological research, or work in analytical or physical chemistry. These students are usually juniors and seniors, but sometimes sophomores and even first-year students will be invited to participate in real laboratory work. They enjoy getting into a working laboratory early in their careers, being useful, and, in some cases, being paid. They report that what they like so much about laboratory chemistry is the combination of thinking and doing, and collaborating in a team-like enterprise where they can learn from one another and share the frustrations and thrills of doing science.

Approaching Biology

Life, in all its variety, is the subject of modern biology. Thus, of all the sciences, biology involves the most facts, the most detail, and, as we have seen in Chapter 2, the most new terms. As in chemistry and physics, there are, of course, important underlying principles and categories into which you will learn to group the facts you are compiling. But, unless you master the full breadth of the subject in all its detail, understanding will elude you.

As in chemistry and physics, part of the subject matter of biology is macroscopic. Organisms, organs, and certain cells, don't have to be imagined; they can be observed in the lab. In other cases, biology is dealing with microscopic entities, which, thanks to modern microscopes, are visible even when they are as small as a single biological molecule. However, biological features, such as chromosomes, genes, and alleles, that can be identified today had to be imagined in the past. The successful student of biology has to learn to imagine as well as to see.

Pamela Parrish is an instructor in introductory biology at the University of Arizona, where she earned a master's degree in toxicology in 1991. She looks forward to a career in managing research in the area of the biology of cancer. She says that biology can be overwhelming for the beginning student, both because of the scope of the material to be covered and because of its detail. General concepts depend on specific examples, and the specifics are best mastered in terms of the concepts. In the beginning, the student has neither. But no one is expected to understand everything the first time it is presented. Successful biology students learn to go back and forth between the concepts and the detail. Sometimes it's wise to focus on the "big picture," the way small variations affect large systems; at other times, it's the small variations that matter. Going back and forth between the whole and the parts is easier if you learn to organize the detail of biology around some unifying concepts.

One unifying concept is *evolution*—the fact that over time and through mutation living species adapt physically to survive. Another is *ecology*—the interaction of organisms and their environments. Once you learn how to think in terms of evolution and ecology, the detail makes more sense and is easier to remember. Several other unifying concepts are discussed below.

Glucose Metabolism

Living things cannot survive without energy for life, but it is not enough simply to ingest food. There has to be a way for an organism to transform that food into energy. Thus, the biochemistry of *metabolism* (the breakdown of materials and the buildup of energy sources out of those materials) is a key concept in the biology of plants, animals, and even bacteria. Glucose metabolism[9] is what makes human activity possible. It's also what makes yeast ferment, producing ethanol—the alcohol in beer. In humans, there are byproducts of metabolism—such as lactic acid (which causes the muscle pain you feel after exercise)—that are produced when food is metabolized in the presence of oxygen. If yeast were metabolized in an environment containing oxygen, it would also produce lactic acid and CO_2. (In some cases, as in the making of cheese and yogurt, certain strains of yeast do produce lactic acid.) The metabolism of glucose in humans (called *respiration*), occurs in an oxygenated environment; the metabolism of glucose in yeast (called *fermentation*), does not. That's why, when we run, we don't produce beer.

About one month into an introductory college biology course, glucose metabolism will be discussed in terms of biochemical pathways. That's because every cell in the body (except the red blood cells) has an energy station where the energy that the cell needs is produced. There are nine main steps in the process of turning glucose into energy—that is, nine sequential chemical reactions, each involving a different *enzyme*.[10] In step one, a phosphate is added; in step two, an enzyme changes the structure of the original glucose molecule; in step three, another phosphate is added in a different place on the glucose molecule. By step four, the glucose molecule has actually split into two separate molecules that differ in the placement of a carbon atom; by step five the two different molecules have become identical. At certain steps involving particular enzymes, not only is the structure of the glucose being changed, but energy in the form of *adenosine triphosphate* (ATP), is being produced. ATP is another molecule, one that is usable by cells, somewhat like the battery in your car. That, in

9. Glucose is not the only substance metabolized in the bodies of animals. Proteins and fats are also made usable by chemical processes, but in this section, only the metabolism of glucose will be discussed. Glucose is found mainly in sugar, fruits, and starchy foods.

10. An enzyme, usually a protein, is a large molecule that facilitates a reaction but doesn't itself change during the reaction. Enzymes affect organic materials; *catalysts* affect inorganic materials in the same way.

fact, is the purpose of the entire nine-step process: to release the energy in glucose so that life in all its functions can continue.

From a biological point of view, starvation occurs either when no glucose is available to produce ATP or when there is a physiological problem that prevents the biochemical reactions from proceeding normally. For example, an enzyme may be lacking, or the glucose may not be able to enter the cell. In the laboratory, it is possible to "starve" cells, by adding a chemical that interferes with ATP production, and then observe the effect of ATP depletion. In biomedical research, scientists may intentionally interfere with cell functioning in order to understand some mechanism or the energy-dependent processes more generally. Certain cells are resistant to certain drugs, for example; the drug gets into the cell but is often pumped out. Since the pumping requires energy, the researcher may want to deplete that cell's energy in order to study what happens when the drug and the cell interact in the absence of energy.

Not all the energy produced through the metabolism of glucose and other food products is used in running and thinking. The body requires energy to maintain itself. In fact, for glucose to arrive at the energy station in the cell, it has to be continuously circulated by a pumping mechanism. Pumping in the body, just like pumping water up from a well, requires the input of energy. In fact, pumping is the way that most molecules get from outside to inside cells. Pumping is also important in maintaining the balance of the system. For example, when salt is in excess, the body pumps it out. When the balance is upset, as with an onslaught of bacteria in so-called "stomach flu," there is too much water in the large intestine, producing diarrhea. In time, a healthy body will overcome the toxins from the bacteria, the intestines will resume pumping, and the excess water will be removed.

Not all food materials can be digested (processed biochemically) by our systems. Celery, for example, consists mainly of cellulose, which is not turned into ATP by the human body. However, cows' digestive processes can turn celery (or grass) into muscle mass (with the help of certain bacteria that live in the cows' intestines); when we eat and digest the meat of a cow, we are getting the energy from grass indirectly.

The process of glucose metabolism is presented early in introductory biology both because it is basic and because it is the oldest system in our bodies. The single-celled organisms from which we evolved had to have glucose metabolism. Four billion years ago, when the earth was new, there was no atmosphere and hence no oxygen. Living organisms that could survive in such an environment (called *anaerobic*) had to be able to process energy in the absence of oxygen. Glucose metabolism is the beginning step for both anaerobic and aerobic living systems. In anaerobic bacteria, one of the products of glucose metabolism is carbon dioxide. Trapped by the earth's gravitational pull, this carbon dioxide built up in the atmosphere and permitted the cells to evolve into more complex systems—plant-like living things—that could utilize carbon dioxide. Plants take in carbon dioxide and give off oxygen during *photosynthesis*. (Plants also respire.) After plants appeared, an oxygenated atmosphere was gradually created; in time, animal-like living things, which require oxygen for their more complex metabolism, could live as well.

What this means is that glucose metabolism is universal. Even anaerobic bacteria go through the same biochemical process of extracting energy from glucose. In the case of anaerobic bacteria, there is a different chemical pathway after step nine because of the absence of oxygen. But under certain conditions we also have to produce energy anaerobically, even though we are air-breathing creatures. For example, during a 100-meter dash, your muscles do not have time to oxygenate, but they are desperately in need of energy. In this situation, the cells in your muscles can temporarily produce energy from glucose without oxygen.

For all these reasons, it is vital to understand glucose metabolism. If you take the time to learn the details of glucose metabolism as presented in lectures and your textbook, you will gain a sense of the physiology of living systems: how they are related to the environment in which they operate (in this case atmospheric) and to evolution itself.

Neurobiology and Behavior

When the cells in our bodies or those of higher animals are in need of energy, a signal is sent to the brain, and we sense "hunger," and instinctively go looking for food. But how do bacteria, which are only one-celled organisms without a nervous system or brain, "know" they're hungry? Biologists are now studying *chemotaxis*, which describes how motile (swimming) bacteria will swim toward higher concentrations of nutrients (sugars and small proteins), and away from higher concentrations of noxious chemicals. Bacteria swim by means of flagella, a very simplified "tail" that propels them forward by the force of a simple motor driven by a gradient of H+ ions across the bacteria membrane. In the absence of chemical signals in their environment, bacteria move in periods of smooth swimming, interrupted by periods of "tumbling" which randomly change their direction. In the presence of nutrients, the tumbling response is suppressed whenever the bacterium is traveling toward higher concentrations of nutrients, and as a result, the smooth, straight, swimming motion brings it closer to food. This behavior is highly adaptive, because those bacteria better equipped to find food in this manner (and stay away from toxins) will survive longer and better.

Scientists are now beginning to believe that when food is unavailable, bacteria have a way of communicating biochemically to other bacteria that it is time to "swarm," that is, to go into a kind of hibernation, since there is not enough food energy available to function at a normal rate. This is amazing in a one-celled organism. Yet it is just as amazing that our multicellular systems can recognize hunger and drive us complex humans to the nearest food store.

That process is part of the systems that are studied in neurobiology and behavior. Although realizing that "I am hungry" and deciding that "It is time to eat" appear to us as abstract thoughts, the methods of communication within our bodies are actually very concrete. Communication signals travel through the body in two ways—electrochemical signals and hormones. The nerves connect the brain and spinal cord to the organs and limbs. Special molecules that reside in nerve cells can transmit signals from one end of the long thin neurons to the other. The cells are lined up end to end in a huge network so that signals can travel through them as through telephone wires. When the signal gets to the end of one nerve cell, chemical *neurotransmitters* are released, which stimulate the next nerve cell to pick up the signal.

Hormones, on the other hand, are transported through the blood to *receptors* in various cells that are specific for each particular hormone.

The human nervous system is not entirely functional at the time of birth. The reason human infants can't walk immediately the way foals and calves can is that the nerves in the lower parts of the infant's body have not yet been sheathed by a membrane called myelin. In the first months of human infancy, myelin is slowly distributed downward over the body. That's why infants' movements appear at first to be spastic; they cannot control arm and leg movement. And that's why babies need diapers; they cannot control their bladders or their bowels. In normal development, myelin will be completely distributed, and full use of the nervous system possible, by the age of three. The disease called multiple sclerosis (MS), tragically reverses this process, causing the retreat of myelin. People who suffer from MS slowly lose nerve control over their limbs and eventually over their entire bodies. Their nerves are still in place, but the myelin that makes it possible to send signals over the nerves has disappeared.

The hormonal system is also not complete at birth, although the organs that produce the hormones are in place. Sex hormones, estrogen (in females) or androgen (in males), are not produced in significant amounts until puberty. These hormones cause an initial growth spurt and then seal in the long bones, which prevents further growth. The later the onset of puberty, the longer the growth period. That's why, in general, girls and boys who develop secondary sexual characteristics early tend to be short.

Hormones are not limited to readying the body for reproduction. They also govern kidney function, growth rate, and metabolism. Certain hormones stimulate the growth of tumors. Some tumors excrete hormones. So in cancer research and cancer therapy, a good deal of attention is being paid to hormones.

Drugs can affect both the nervous and the hormonal systems. They interfere with neurotransmission either by preventing the release of a neurotransmitter or by blocking the receptor on the next nerve ending. They can also affect the glands that produce hormones. The craving for a drug can be physiological, meaning one is physically dependent and feels sick when the drug is withdrawn. When drug need is physiological, in time the body's craving overrides all else. In the case of rats that have been forcibly addicted to cocaine, for example, researchers find that a rat will do anything, even inflict pain on itself, to get more of the drug.

Learning and Memory

The brain is the least understood part of the body in humans and animals, because both of its physiology and its psychology. In fact, the brain and our environment are mutually interactive. Animals, like plants, are extremely responsive to the stimuli provided by their surroundings. In rats deprived of light, the *dendrites* (little spines on the nerve cells of the optic nerve) will remain underdeveloped. But when they are finally placed in light, they develop additional spines on those cells.

We humans have a built-in capacity for language even before we learn to speak through mimicry. Physiological linguists believe that a readiness for vocabulary and even for grammar is part of human "hardware." Recall the story of Tarzan, which is fictional but biologically sound. As a child, Tarzan did not develop anything more than grunting behavior in the community of apes in which he was brought up. But unlike the other apes, he could learn language when he met another human.

How do we learn, from a biological point of view? One form of learning is training. We train the nerve signals to take certain pathways by way of *habituation* and *sensitization*. This training, which is one form of learning, involves memory and occurs at the level of the *synapse*. (The synapse used to be thought of as the gap between nerve endings, where the neurotransmitters operate. Now neurobiologists define the synapse as the entire structure at nerve endings that permits the transfer of signals.) Initially, any new *input*, whether a piece of information or a skill one is learning, produces a response called an *arousal*, which can be pleasurable or not pleasurable. When that input is repeated many times, the nerve becomes habituated and ignores the arousal. From a physiological point of view, the new input no longer "bothers" the system; it feels "natural." The feeling you get is that you know how to do something you did not know how to do before.

Where in the brain are memories stored? Mary Kennedy, a professor of neurobiology and behavior at the California Institute of Technology, is studying the biochemistry of memory storage. She believes that different kinds of memories are stored in many different places in the brain: visual memories in the brain's visual areas, motor memories in motor areas, and so on. Complex memories, the kind you need to store in order to pass biology in college, are most probably dispersed throughout areas of the brain called the *association cortex*. However, one brain structure, which lies just beneath the cortex, seems to play a particularly important role in the formation of memories. Its name, the *hippocampus*, comes from the Greek word for sea horse, which some early anatomists thought it resembled.[11] Kennedy and her colleagues believe that the physical changes in the *neurons* (nerve cells) of the hippocampus occur as a result of electrochemical activity.[12] They believe that the neuron integrates information coming from its many synapses by totaling the electric signals produced by each of them. If, for example, many synapses fire at once (the information is coming in from a variety of sources) or one or two synapses fire many times very rapidly (you're repeating the exercise many times), longer-term chemical changes occur. Kennedy and her colleagues believe that these longer-term changes are the biochemical basis of memory.

There are three kinds of memory banks in the brain:

1. Procedural memory involves only neuron responses, which permit us to ride our bikes, for example, without paying much attention to what we are doing unless our attention is suddenly drawn to a problem such as a near collision.

2. Short-term memory permits us to capture and repeat small amounts of information fast. But the information (as discovered by every student who waits until the last minute to cram for an exam) is not retained very long. The disadvantage of short-term memory is just that: it is short-term. The advantage is that it is easy and extremely fast.

3. Longer term memory is what biologists call stable memory. It is harder to get information into long-term memory because the process evidently involves the chemical synthesis of proteins, not just neuron responses. This means that it will take longer, physiologically, to make the changes that are required to hold things in memory. On the other hand, because new molecules are created, the information is more permanent, often lasting even through concussions, electric shock, and other traumas.

11. Mary B. Kennedy, "How does the Brain Learn?" *Engineering and Science* (September 1990), 4 ff.

12. In the laboratory, they insert microelectrodes into neurons and find that there is a difference in voltage between the inside and the outside of a neuron. This creates a potential difference along the synapse, which in turn allows sodium, potassium, and calcium ions to enter the cell.

Using short-term memory, that is, memorizing facts, is obviously not the way to succeed in learning any subject. Studying in ways that engage long-term memory is a much better investment of your time. Most of all, you will want to develop the reasoning skills that you will need in science. This will permit you to anticipate next steps, organize material in the most efficient ways, and, most important, apply information so that you can solve problems, construct descriptive sentences, and draw conclusions. This is what biologists mean when they say that learning requires not only neuron response, but alteration of behavior.

Adaptation and Evolution

Learning is a survival mechanism in animals; they do not function on instinct alone. Animals will learn to recognize a poisonous plant from experience (if it's not so toxic that it kills them the first time), and rats will learn to find their way out of a maze if they are given the proper stimulus to learn. Humans also learn from experience and from the transmission of other humans' experience through parental guidance and schooling. Much of learning is adaptation to variations in the environment. When a single animal alters its behavior to better accommodate itself to a certain surrounding, it has a better chance of surviving as an individual to reproduce. When we speak of a species surviving—Darwin's "survival of the fittest"—we are talking about much more than individual adaptation. Survival at the species level is a matter of evolution based not on learning, but rather on certain natural (mutational) variations in the species that are favored because they are advantageous.

Even though humans are considered to be a single species, there are variations among human races that show long-term adaptation to different environments. Groups that developed in Africa, which is generally warm and sunny, could get ample supplies of vitamin D from exposure to the sun, and thus did not need to be able to digest milk beyond infancy. In the northernmost parts of Europe, however, where sunshine is scarce, people had to get vitamin D in other ways. Hence, they became cattle-raisers, and those with sufficient lactose enzymes to digest milk (even into adulthood) were favored by evolution. Blacks today very often show lactose intolerance, which does not interfere with their fitness to survive. Similarly, the descendants of people who survived in Africa also show certain abnormalities of their hemoglobin, which affect the distribution of oxygen to their cells. This deficiency is called sickle-cell anemia. Although this is a handicap, it also protects against malaria, which is why it was favored in evolution. In certain parts of Africa, 40 percent of the population now carry the sickle-cell trait, but only 10 percent of African-Americans have it. This suggests that adaptation on an individual level can take place more quickly than species evolution.

Many different kinds of biologists study neurobiology and behavior. Wildlife biologists travel to the field, the habitats where certain animals reside, and study their behavior. For example, Jane Goodall is a primatologist who spent her life observing chimpanzees. In laboratories, other biologists are trying to understand the biochemical and other mechanisms that account both for behavior and for change of behavior. Cognitive scientists study learning in higher species, including our own. Genetics is important in understanding behavior, since survival in evolutionary terms means passing on the better-adapted genes to offspring. As in all fields of biology, it takes many different specialists to gather the details, but, to make sense of their research, all of them keep their eyes on the basic issues.

The Biology Lab

The purposes of the biology lab are not very different from those of chemistry and physics labs. Your professor wants you to experience the scientific process as soon as possible. And so you will be shown how to set up an experiment; observe; take measurements; and collect, analyze, and draw conclusions from your data. But at least as important as the development of your technical skills is the deepening of your understanding of the concepts that are being introduced in lecture. In Mendelian genetics, for example, you are told that mating in the biological world is probabilistic within species. When you are in the lab pulling beans of different colors out of large bags of beans, you see mathematical probabilities in action.

Reproduction also becomes less mysterious in the lab when you can watch the reproduction of sea urchins. First, you make the sea urchins shed their sex cells or *gametes* (sperm, in the male;

eggs, in the female). Then, under a microscope, using a pipette, you collect a small supply of sperm and fertilize the sea urchin eggs. Within minutes, you can see the fertilized eggs begin to divide; within two hours, a number of divisions take place. All this makes the complex process of development very graphic.

Fermentation can be observed in the lab by combining sugar, yeast, and a little water, and watching the yeast ferment. It is obvious in the lab, if it wasn't in lecture, that the yeast is drawing energy from the sugar, and that byproducts of CO_2 and ethanol are being produced. Students can actually detect the presence of CO_2 as a product of fermentation in the lab. In the case of plant photosynthesis, students can show that chlorophyll and light energy are essential to the production of sugars and/or starches.

Laboratories are also used in biology to demonstrate the higher, more complicated techniques professionals employ in their research. Using protein gel *electrophoresis* (separating proteins in a jello-like medium by applying a potential difference, i.e., voltage), students can see how proteins, amino acids, and nucleic acids can be isolated and identified. Genetic engineering techniques can also be demonstrated in the lab. Different cultures of bacteria are grown on different plates, and foreign DNA are added to some. The experimenter can see if those bacteria containing the transplanted gene are growing a new form.

Finally, the laboratory provides invaluable opportunities to visualize the structures and some of the processes of living systems. Cells are presented in lecture in neatly drawn pictures, but when students look at real cells under a microscope, they appear far more complicated. The casing of the cell, the cell membrane, appears darker because the walls are larger than the other components of the cell, its *organelles*. You can usually locate the nucleus of a cell, even though it is not always near the cell's center. Cells come in many different shapes and sizes, but most of the cells from the same tissue show the same pattern of structure, such as columns or cubes. Using a microscope to look at cells from the same and different tissues, with their assortment of shapes and arrangements, drives home the point of developmental differentiation.

By the time you finish the cell work of your introductory biology lab, you will no longer be in awe of the trained pathologist who can put a "mystery tissue" under a microscope and tell you

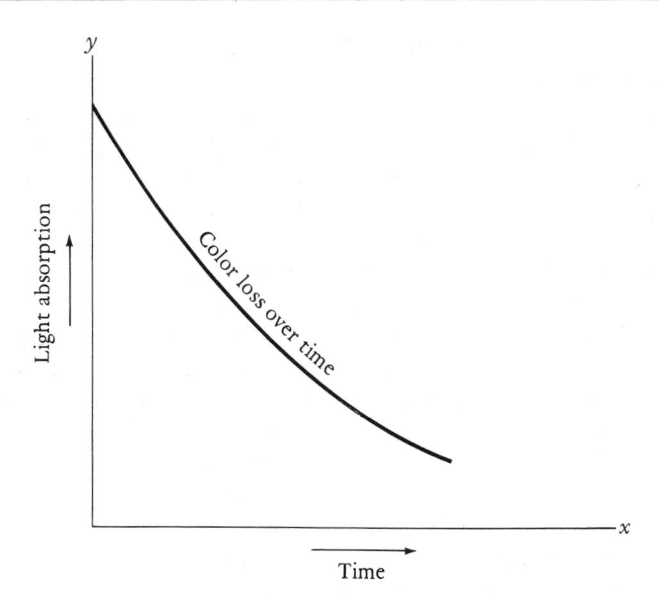

FIGURE 3.12
The relationship between time and light absorption.

what it is, where it comes from, and what it does! You, too, will be able to read the clues.

When collecting data in a laboratory situation, it is important to arrange your observations in such a way that you can make sense of them. For example, in experiments having to do with chlorophyll, the amount of color being removed can be measured with a certain instrument that will give you a numerical value for the amount of light that particular color is absorbing. A graph that compares such values over a certain time period provides a visual aid for remembering the relationship between time and light absorption. The graph in Figure 3.12 shows that change in the absorption is best represented by a curve where most of the color is lost at the beginning and then the loss drops off.

What Is "Mastery" in Biology?

Seven years ago Pamela Parrish was a beginning student in biology. Today she has her master's degree in toxicology and has gained much confidence in her ability to learn biology. Although she doesn't know everything there is to know about living systems—no one does—she has the kind of understanding that permits her to find out more. Faced with an unknown substance, for example, she can figure out, at the very minimum, what it is *not*. Then she can proceed, by means of her knowledge of biology and chemistry, to subject it to certain tests.

If she sees an outer membrane, she can assume the substance is a cell. If she stains it with a dye, she is able to see certain structures or characteristics, such as its organelles. Now she can begin to compare and contrast it to other cells she knows more about. In time, she can unravel the mystery and, like a pathologist, make a reasonable guess as to what it must be.

Having this level of understanding of biology gives Parrish both the confidence and the ability to make independent judgments, and thus to be skeptical about "preliminary observations" that make their way into the popular press. She has an understanding of what makes a study valid, and how many trials and retrials are needed before a scientist can be certain that he or she has found a cause to explain an effect.

After seven years of mastering pieces of the vast puzzle that is called biology, Parrish is well on her way to having the pieces in place. The frustration of learning biology, she says, is that there are so many pieces. Because life systems are designed to ensure survival, they are *redundant*, that is many different processes are available for the living system to employ, so that the failure of one does not necessarily lead to death. Still, there is a "big picture," and the joy in studying biology, says Parrish, is getting the whole picture put together. Even though it was hard for her, as it will be for any beginning student, she kept at it because the whole was so interesting and the payoff so rewarding in terms of both the professional career it opened up for her and in terms of her own personal science literacy.

The Reality of Science

Understanding science involves not only laboratory work, but problem solving, and making connections among concepts. Although there are many different terms to learn, Charles Holbrow, a distinguished professor of physics at Colgate University, points out that knowing the names of things in science is not enough.[13]

The reality of science is nature itself: how forces can be described, and how they affect material objects; what matter is made of, and how elements and their compounds react; the processes of life forms—their use of energy, their ability to repair themselves, and their survival. But at least as important in understanding science is another reality, the models scientists create and employ to make sense of the natural world. "Without models," writes Arthur Young, a professor of physics and astronomy at San Diego State University, "scientific inquiry would be in jeopardy of becoming aimless. With them, comes a conceptual filter which directs further observation and discriminates that which is important from that which is not."[14]

In the process of learning science, you will be exposed to many very important models, which have been called the structure of scientific thought. It is your challenge (and that of your professors) to learn the difference, as Young expresses it, "between what is real and what is conceived, i.e., between the data and the models." Chapter 4, "Mathematics, Models, and Measurement," describes the process of scientific knowing, and shows you why the data scientists collect from their experiments and observation are so important—yet insufficient in themselves—for scientific inquiry.

13. Review of Arnold Arons, *Teaching Introductory Physics*, in *Physics Today* (December 1990): p. 68.

14. Arthur Young, "Science in the Liberal Arts Curriculum: A Personal View," in *Publications of the Astronomical Society of the Pacific*, (December 1983): p. 941.

4

Mathematics, Models, and Measurement

One of the challenges of learning science is to balance the accumulation of new knowledge with the acquisition of new skills. As we noted in Chapter 3, when we compared learning physics to learning to swim, science is a subject in which you have to participate in order to succeed. This means practice as well as study; training as well as education. Introductory science will teach you techniques for doing scientific work and introduce you to some of the most exciting and far-reaching ideas in the history of Western civilization, such as Newton's three laws of motion and Avogadro's insight.

The skills you will learn in science are as empowering as they are indispensable. This is what Francis Bacon (1561-1626) meant when he said, "The scientific method allows ordinary people to do extraordinary things." Science can be done by almost anyone who has the curiosity, the persistence, and the training to do the work. The notion that only geniuses do science misses this point.

The Scientific Method

What, then, is the scientific method? How do scientists define the problems they will study? How do they decide what to do and in what order? How do they protect themselves from error? Science, it is said, has three distinguishing attributes:

1. Science is *objective*, meaning that no matter who repeats an investigation, the result will always be the same.

2. Science is *empirical*, meaning that its findings are measurable and that every theory can be tested. No scientific theory can be *proved*, but a great deal of evidence (including predictions) can be accumulated in its favor.

3. Complicated problems in science are best broken into smaller problems and solved, the assumption being that the whole is the sum of its parts.

The corresponding philosophical terms are: *value neutrality* for objective, *positivism* for empirical, and *reductionism* for the belief that large scientific problems are divisible into smaller ones. You won't encounter these terms unless you study the history and/or philosophy of science, but you will see value neutrality, positivism, and reductionism in action as you begin to study science.

This chapter shows you how mathematics, models, and measurement serve science in achieving its goals: to describe and predict the outcome of natural phenomena. Because science is empirical, it begins with observation and measurement. Because measurement is quantitative, science is inevitably mathematical. But observation, no matter how accurate, and measurement, no matter how exact, are not enough. Because scientists need to make sense of their observations and measurements, they construct models: of the structure of matter, of the solar system, of evolution, and even of the elemental atom. Their models provide them with a way of thinking about the whole before the whole comes entirely into view.

You do not need to understand the philosophy of science to learn science. Many working scientists do not see any need to discuss either the methods they rely on in their work or the philosophical justification for what they do in the field or in the lab. The American physicist, Percy Bridgman (1882-1961), spoke for many when he said, "The scientific method is what working scientists do, not what other people or even scientists themselves may say about it."[1] Still, we believe that an understanding of the structure of science will enhance your appreciation and contribute to your success in learning science. Throughout your involvement in science, remember that it is an empirical subject. The quality of observation is dependent on the accuracy of measurement, and the usefulness of both is dependent on the model.

Mathematics—Quantitative Thinking

One way to appreciate the power of quantitative thinking in science is to try to imagine what science would be like without it. Copernicus is credited with proposing a heliocentric model, that the earth and the other planets circle the sun, rather than that the sun and planets circle the earth. But Copernicus's model was oversimplified. He believed incorrectly that planetary orbits were circular because his measurements were not precise. The only technique available to Copernicus to measure the positions of the planets was naked-eye sighting with the aid of a primitive aiming device. The angular distance between a fixed star and Mars, for example, gave him measurements accurate only to within one-fifth of a degree (12 minutes of arc). As a result of this imprecision, Copernicus never realized that planets do not move in circles. Although his heliocentric theory was revolutionary, he followed the conventional Greek notion of the circle as a "heavenly form of motion." To the end of his life, as a result of both inaccuracies of measurement and his belief that the spherical shape of the planets caused their circular motion, Copernicus never discovered elliptical orbits.

Sixty years later, another astronomer, Johannes Kepler, was able to work with much better observational data, recorded by Tycho Brahe. These measurements were 20 times more accurate than those available to Copernicus, and the best available until the telescope was invented. Kepler had so much confidence in Brahe (and in Copernicus) that he struggled for four years to fit Brahe's data on Mars to a circular orbit of any kind. But, however hard he tried, his best circular orbit had a discrepancy with Brahe's

1. Percy W. Bridgman, *Reflections of a Physicist*, 2nd Edition (New York, Philosophical Library, 1955), pp. 81-83.

data by about one-eighth of a degree (8 minutes of arc). In the face of this stubborn quantitative fact, Kepler tried various other circle-like shapes until he had to conclude that the elliptical orbit was the only one that fit the data.[2]

Kepler's Laws

Kepler is justly famous for the three laws of planetary motion which accurately describe the behavior of all nine planets in our solar system—only six planets were known in Kepler's time—as well as thousands of asteroids and comets that have been discovered since. The three laws are not laws in the same way that Newton's laws of motion are; rather, each of Kepler's laws is considered by scientists to be a mathematical summary of measurements. Suppose we eliminated the mathematical aspects from Kepler's three laws. How much would they tell us? How useful would they be? Here are Kepler's three laws, first stated in nonquantitative terms, then as Kepler formulated them. Note how Kepler's laws allow us to do specific calculations.

1. Planets orbit around the sun. (Kepler: Planets have elliptical orbits with the sun at one of their focuses.)

2. A planet moves faster when it is close to the sun than when it is far away from it. (Kepler: The line connecting the sun and the planet sweeps equal areas in equal times.)

3. The planet with a larger orbital radius takes a longer time to do a complete revolution around the sun. (Kepler: The square of the period of revolution [T] of the planet is proportional to the cube of the average distance [R] between the sun and the planet.)

   The third law is expressed in mathematical notation as

$$T^2 = KR^3$$

2. This analysis is much indebted to the chapter on Kepler's Laws in Gerald Holton and Duane Roller, *Foundations of Physical Science* (Reading, Mass., Addison-Wesley, 1958).

where T is the period of planetary motion around the sun, R is the average distance (over the ellipse) between the planet and the sun, and K is a constant of proportionality.

Kepler's mathematical formulation not only summarized all the astronomical measurements having to do with planetary motion but, according to historians of science, helped establish the equation as the prototypical form of laws in physical science. Previous to Kepler, mathematicians had used equations, but scientists had not yet realized the power of mathematical equations; no wonder Kepler is regarded as the first modern scientist.

Without Kepler's laws, it is unlikely that Newton would have been able to confirm as convincingly as he did the theory of universal gravitation. It was Newton's genius to discover, half a century after Kepler's three laws were published, that the inverse square force that worked on earth explained two of Kepler's laws, the first and the third.[3] As noted in Chapter 2, Newton's inverse square law is:

$$F = G\frac{m_1 m_2}{r^2}$$

where F is the force of attraction, G is the universal gravitational constant, m_1 is the mass of the sun, m_2 is the mass of a planet, and r, the distance between them.

Calculus

It is no accident that Isaac Newton, in his pursuit of universal laws of motion, also invented calculus. After Descartes, mathematicians could plot functions on a graph, as you have learned to do already in high school. What they needed was a way to find the slope of a function from a given curve. The mathematics of slope finding was the first task of the new mathematics called calculus. Descartes struggled with this problem without solving it. Newton, pursuing many interesting unsolved problems in mathematics as well as physics, realized that what mathematics needed

3. Newton also proved geometrically that Kepler's second law comes as the result of any force that is directed toward a fixed center.

was a means of expressing the infinitely small. Until then, algebra and geometry had given scientists many and varied means of describing a quantity that changes in response to a change in something else, but they couldn't describe the rate at which those changes were taking place in the dependent quantity. They could find the slope of a curve geometrically by "eyeballing" the plot (Figure 4.1), but they could not find an algebraic expression for the rate at which the curve was changing.

Nature is always changing, or in flux; indeed, Newton's name for calculus was "fluxions," because the power of his new mathematical tool was that it could capture in a quantitative form the patterns of change in nature.

When you learn how to calculate *derivatives* of *functions* in your first course in calculus, the laws of physics will appear much more straightforward and understandable than if you have to try to make sense of them using algebra. This is why calculus-based "University Physics" courses are better and actually easier to understand (although that is not their reputation) than "College Physics," which is taught without calculus. Describing rates of change in velocity or position without calculus is both more awkward and less precise.

In daily life people usually think about speed in terms of average speed, calculated as a simple ratio: the distance traveled within a certain time (speed = distance/time). But when a more precise measure of speed is needed, as in physics, an average will not do. To find a speed at a particular moment, it is possible, using calculus, to begin with an average speed over a short time interval, then to shrink that time interval to zero, and then to derive an exact mathematical expression for *instantaneous speed.*

In algebra, if you know the coefficients a and b in the equation $ax + b = 0$, you can find the unknown x. In calculus, if you know the function (in this case, displacement as a function of time), you can find a derivative, the slope of the curve (in this case, the instantaneous speed as a function of time). Conversely, if you know the derivative, you can find the original function. Finding a derivative from a given function is called *differentiation* in calculus; finding the function from its derivative is called *integration.*

The mathematical concept of a function is not limited to physics. In most of science, quantities depend on other quantities, or on conditions expressed as quantities. Bacteria multiply as a

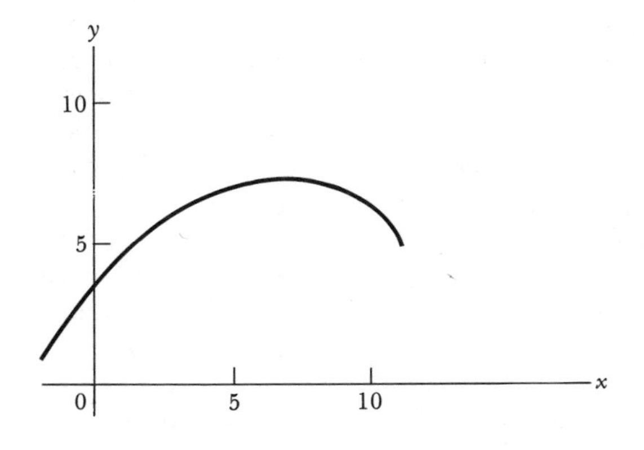

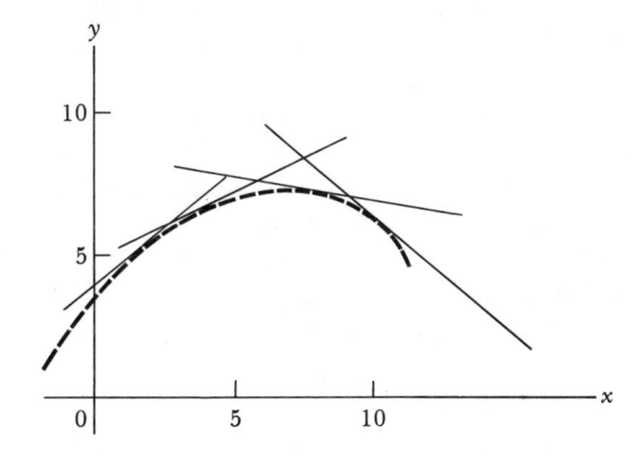

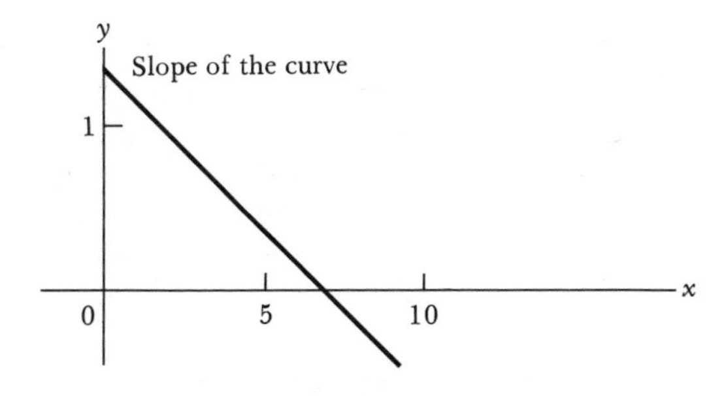

FIGURE 4.1
Eyeballing to find the slope.

function of time. The rates of chemical reactions in general are functions of temperature and concentration. The expression of relationships among quantities as functions is a common application of mathematics to science including social science. Price, as you will learn in economics, is a function of two quantities: supply and demand. The dependence of one quantity (x) on another (t) can be expressed in mathematical notation as $x = f(t)$. But if you want to find the rate of change of x as t changes, you have to use calculus.

The way calculus is applied in physics is as follows. The quantity acceleration (a) (as in Newton's second law of motion, $F = ma$) is defined in physics as "the second derivative of position (displacement) with respect to time." The reason acceleration is the second and not just the derivative of position with respect to time is that acceleration is a rate of change in velocity, and velocity, in turn, is the rate of change of position. Hence, acceleration is a kind of "double derivative" called mathematically the second derivative.

When a function is known, you can find a derivative; when a derivative is known, you can find a function. But when neither the function nor its derivative is known, one is obliged to resort to more advanced techniques, such as *differential equations*. For example, when F in $F = ma$ is constant (when the force is unchanging with time), and m (mass) is also constant, finding displacement as a function of time $x(t)$ from $F = ma$ is a straightforward problem in integration. But if F is a combination of several forces, in which one force, for example, depends on the velocity of the object and another force depends on the position of the object (as in a problem involving a damped spring), simple integration is not enough to find what kind of motion results from these forces, and $F = ma$ becomes a differential equation.

Because so many of the processes in nature are the result of combinations of changes, most natural phenomena can be expressed in the form of a differential equation. If there is more than one parameter, such as when pressure and temperature both affect volume, the differential equation will be more complicated (these are called partial differential equations), but it is a differential equation nonetheless.

Some natural phenomena, however, such as the relationship between the holding capacity of a rain forest and the flow rate of a major river, depend on the cumulative effect of past events, such as rainfall in the watershed. Problems like this one are best understood not as differential equations but as integral equations. (Integration gives the area under the curve; the accumulated change.)

In most real instances, natural phenomena present too many interrelated factors to deal with at once. To cope with them, scientists may decide to simplify the problem by intentionally overlooking certain factors. In the river example above, direct evaporation of rain into water vapor might be ignored; or instead of representing rainfall precisely as a function of time, one might use, instead, monthly averages. The advantage of simplification is that the scientist can do a "first pass" at a problem. For problems like these, the computer's ability to handle rapid numerical calculations makes it an invaluable tool.

Logarithms

Not all the mathematical techniques used by scientists are advanced. Some, like the logarithms and exponentials you first encountered in high school, are not complicated, but contribute substantially to the way scientists not only do their calculations but think about and describe natural phenomena.

Logarithms, you will recall, are numbers expressed as exponents of 10. The log of 100 equals 2; the log of 1,000 is 3; and the log of 500 is 2.7. In the "dark ages" (that is, before handheld calculators), students would refer to a table of logarithms as a way of avoiding lengthy multiplication and division problems. For example, instead of multiplying 638 by 219, you can add the log of 638 (2.805), and the log of 219 (2.34) and get a total of 5.145. Referring back to the table of logarithms, you find that 5.145 is the log of 139,722, which is the product of 638 and 219. The slide rule, once the first purchase of beginning science and engineering students (now but a museum piece), is a ruler on a logarithmic scale. Adding two lengths on a slide rule, each corresponding to the logarithm of a particular number, is equivalent to multiplying those numbers. You can see why, before calculators, the slide rule was the scientists' and engineers' best friend.

But the scientific calculator hasn't put logarithms entirely out of business. Logarithms have other uses in science. There are quantities in chemistry, physics, and astronomy that are conveniently scaled by logarithms. These are quantities which have very large variations. It is not "wrong" but it is certainly not convenient to describe the brightness of heavenly bodies, for example, on a linear scale where the brightest, the sun, would be billions of times brighter than the faintest galaxy, barely visible through the most powerful telescopes. Instead, astronomers use a modified negative logarithmic scale. On this scale, the sun is "magnitude minus 27"; the faintest detectable galaxy might correspond to "magnitude plus 25." Because of the compression of logarithmic scaling, these numbers are easier for astronomers to deal with—the difference in magnitude of two stars is 2.5 times the log of the ratio of brightness of the two stars.

The same is true of hydrogen ion concentrations in chemistry. What is commonly known as the *pH*—a measure of acidity in a solution—is also based on a negative logarithm. For pure water, at 25 degrees Celsius, the concentration of H^+ (hydrogen ion)[4] is 1×10^{-7}. Since the logarithm of this number is -7, the pH, which is defined as "the negative logarithm of the hydrogen ion concentration" is, in this instance, 7. That is, the pH of water is 7. The notation pH derives from the French *pouvoir hydrogene,* meaning the power of hydrogen. Quantitatively speaking, you can remember this as the *exponential power* of hydrogen concentration, which will remind you to employ a log scale when you are doing pH calculations.

It helps you judge the acidity or alkalinity of many things in nature. Any substance whose pH is below 7 is acid; if its pH is above 7, it is basic (or alkaline). Thus, vinegar has a pH of 5.2 (moderately acidic), and household ammonia has a pH of 10. The pH of your blood is normally 7.4, which is a lot easier than saying it has a hydrogen ion concentration of 4×10^{-8} M (where M stands for moles per liter or *molarity*).

Sometimes, in solving problems in science, what is missing is the exponential factor—that is, the power to which other elements in the equation are raised. When the exponential is the unknown, scientists very often convert it to the logarithm of the equation, instead of trying to solve the equation directly. In this way, you can turn an unsolvable problem into a solvable one—a straightforward linear equation. In the case of radioactive decay, for example, the rate of decay is expressed by the equation

$$N = N_0 e^{-\lambda t}$$

4. Some chemistry textbooks identify the hydrogen ion as H_3O^+ instead of H^+.

where N is the number of radioactive nuclei remaining at time t, N_0 is the number present at the beginning of the measurement (when $t = 0$), t is elapsed time, and λ is the decay constant. (e is the basis of natural logarithms as 10 is the basis of common logarithms, and is equal to 2.7183.)

This expression of radioactive decay is a good example of a solution of a differential equation. The number of nuclei that decay at a given moment is proportional to those that are left (N). If the proportionality constant is (λ), this can be expressed as:

$$dN/dt = -\lambda N$$

where dN/dt is the rate of decay. This is the differential equation for radioactive decay. The negative sign in front of λN signifies that the rate is negative because nuclei are being removed.

By taking the logarithm of both sides of the original decay expression (not of the differential equation), the unknown quantity, λ, becomes just a slope of a straight line, the m of the equation $y = mx + b$.

When you get to these kinds of measurements in the laboratory, if computers are not available, you may be asked to plot them on semilog paper. Log paper comes in different cycles and configurations, depending on the researchers' needs. Log-log paper has a log scale on both the vertical and the horizontal axes. (See Figure 4.2.) Semilog paper provides the user with log scale on the vertical axis and linear scale on the horizontal axis. (See Figure 4.3.) If the range of variation in a particular measurement is, say, 1 to 100, that data can be plotted on 2-cycle semilog paper where log of 1 is 0 and log of 100 is 2. Raw data, however, rarely start with 1 and end with 100, so the researcher prefers 3-cycle log paper (2 plus 1) for this range. If the range of variation is one million, say, from 3 to 2×10^6, then 7-cycle log paper $(6 + 1)$ will be selected.

One example of logarithmic analysis of data is the study of the relation of (human) age to the incidence of cancer. We all know that cancer is more common in older people; but why? Physicist Herbert Holomon speculated that the appearance of a cancer might be like a *spontaneous nucleation*, a process similar to the formation of a raindrop out of water vapor. One characteristic

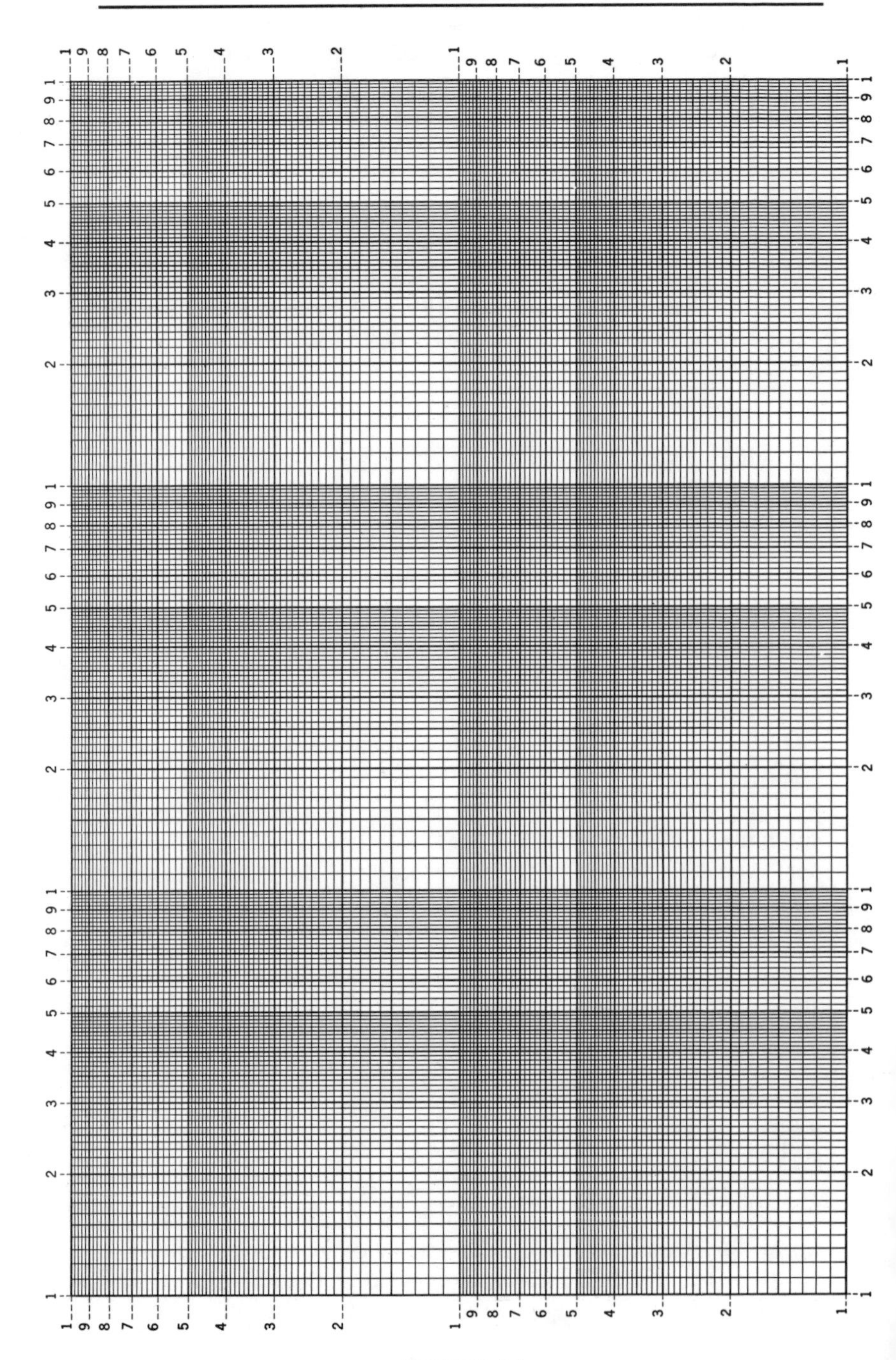

FIGURE 4.2
Example of log-log paper.

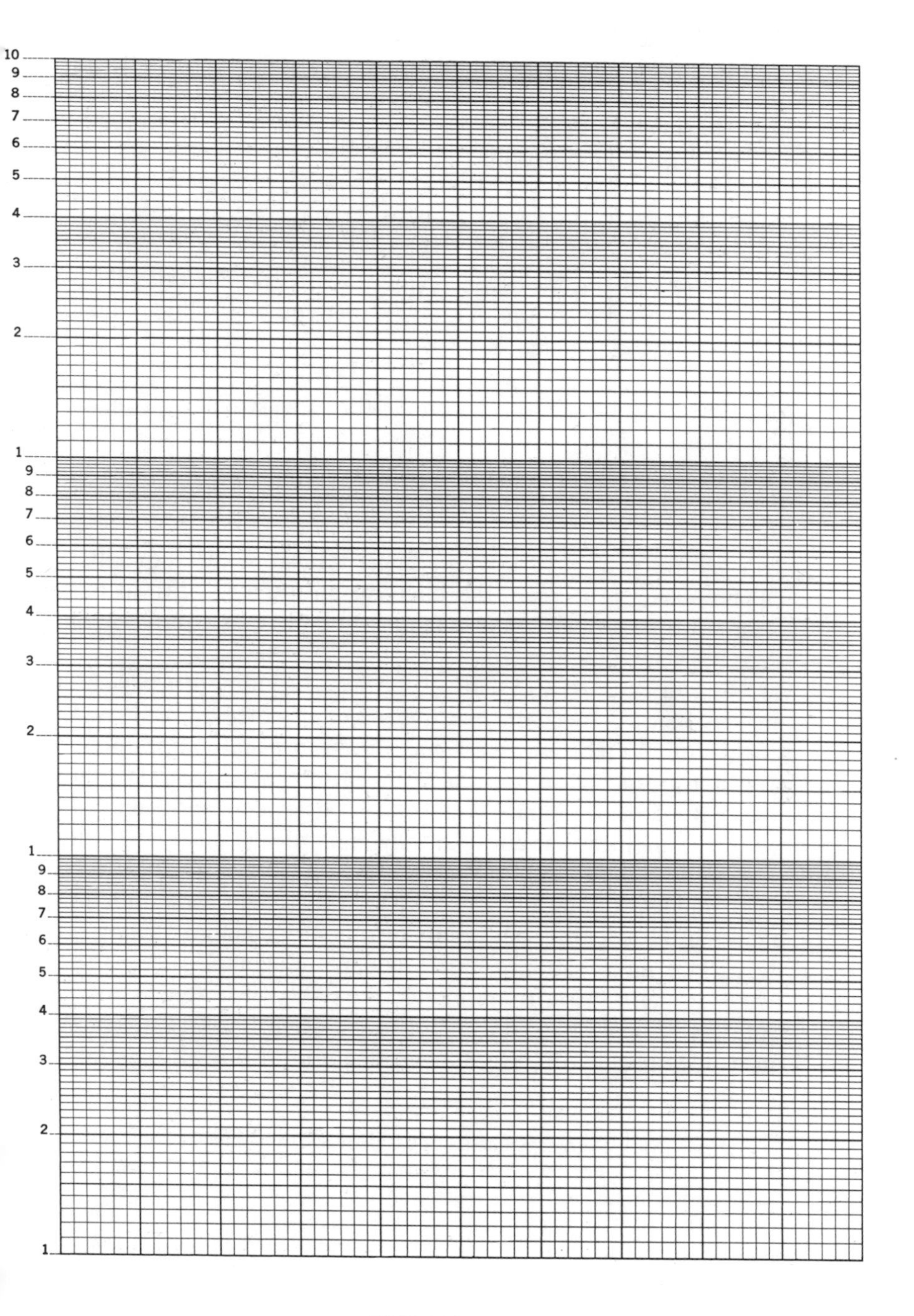

FIGURE 4.3
Example of semilog paper.

of a nucleation process is that it depends on the cube of elapsed time (t^3 where t is time). He did a search for data and found that, in fact, n in t^n for incidence of cancer was close to 3. Here is how log-log paper comes in.

Start with the function:

$$a = Bt^n$$

where a is the incidence of the disease and t is the age in years. B and n are the unknowns that you wish to determine by studying the clinical data.

When you plot these data on log-log paper, you are changing the relationship to:

$$\log a = \log B + n \log t$$

This permits you to work with a linear instead of an exponential equation. Log B becomes the intercept on the y axis just as the term b in $y = mx + b$, and n is now nothing more complicated than the slope of a line.[5]

As you can see, the tools of science—including log paper—are designed to meet the needs of scientific measurement and calculation. As technology advances and more and more sophisticated instrumentation becomes available, the older forms fall out of use. But unless you have your measuring device directly hooked up to a computer that automatically analyzes the data, you will have to do some of this yourself.

If you use log-log and semilog paper, you might want to think about all the scientists who came before you, taking data, laying it out, looking for patterns, thinking new thoughts. It is probable that Marie and Pierre Curie plotted the radioactive decay curves from the sample of radium they had extracted from pitchblende on semilog paper. We have some evidence for this because in her

5. Herbert Holomon's career exhibits the flexibility and intellectual breadth available to a scientist. He worked as a research scientist at General Electric, was Undersecretary of Commerce in the Kennedy administration, and later president of Oklahoma State University. He never worked directly in cancer research but was able to bring knowledge from another area to bear on cancer-related questions.

textbook, *Radioactivité*, Marie Curie illustrates with a semilog plot how one can determine two different decay constants from a single decay curve. The reason these calculations were so crucial in her experiments is that radium decays into radon at a certain rate, which in turn, decays at a different rate—a typically complex decay scheme in natural radioactive substances.

MODELS

There is more than one meaning of the term model in natural science. When biologists talk about a mouse model, they mean that they are using a mouse in place of a human being to study the effect of a certain agent on a biological system. What chemists and physicists generally mean by model is a two- or three-dimensional "picture" of some natural object, such as a molecule. Chemists and physicists also model complex phenomena. The liquid-drop model was used to represent atomic nuclei. One can more easily describe the behavior of a droplet of water than the collective behavior of neutrons and protons which make up an atomic nucleus. The liquid-drop model can even enable one to describe the fission of a uranium nucleus. What happens as the drop absorbs an extra droplet, like a neutron, is that the surface tension can no longer hold the drop together and the drop, that is, the nucleus, divides into two nearly equal parts.

Simulation

Lately, scientists and engineers have learned to simulate natural events by means of computers. For example, when trying to design a new airplane wing, engineers construct a simulation of a wind-tunnel experiment mathematically, and then test their design on this simulation before going to the expense of actual wind-tunnel testing. Even the design and improvement of new computer chips are first simulated on computers. Modeling helps scientists understand the behavior of complex systems so that, for example, they can better make predictions about weather, or better understand inaccessible phenomena such as events in astronomy. Computer simulations of experiments have become so commonplace that some people refer to the results obtained by computer modeling as "soft data," as compared to the "hard data" they get in a real experimental laboratory.

Historically, models have played a very important part both in accelerating and retarding scientific discovery. Thomas Kuhn, a historian of science, says science proceeds by models (he calls them *paradigms*), which scientists create as a way of making sense of their observations and experiments.[6] Once a good model is established, however, it is hard to overturn because (1) it explains the phenomena so well, (2) it is usually a vast improvement over the past model, and (3) the dominant paradigm, as Kuhn calls it, becomes part of mainstream science, as represented, for example, in the standard textbooks. Inevitably, even the best models may stand in the way of new knowledge. To the extent that the model answers all the old questions, new questions may be discouraged.

In the eighteenth century, for example, the so-called "phlogiston theory" stood in the way of Lavoisier's efforts to understand the chemistry of combustion and the formation of rust (both of which would turn out to be forms of oxidation). Every combustible substance was thought to contain a mythical element, phlogiston. When the substance burned, phlogiston was released, leaving ashes. The theory was successfully overturned by Lavoisier's discovery of oxygen as the element involved in combustion. This put phlogiston to rest. But later, Lavoisier's own "caloric theory"—that heat is a material fluid—needed to be overturned as well and replaced by the first law of thermodynamics, which states that heat is a form of energy (see Chapter 1). Just as in prior centuries, new models continually replace older models in chemistry, physics, and biology.

In Kuhn's view, the Copernican revolution was a "paradigm-shift," one major change in the fundamental model of the universe, but by no means the only one. Copernicus realized that the earth was not the center of the universe, but had no idea how vast the universe was. Today, astronomers view our solar system as but a speck near the edge of one of the millions of galaxies. What is debated today is not the model of the universe, but whether the Big Bang model explains everything about its origin.

6. Thomas Kuhn, *The Structure of Scientific Revolution* (Chicago, University of Chicago Press, 1962).

Models in Chemistry

Of all the branches of natural science, chemistry is rooted in a powerful—meaning dynamic and productive—model, the atomic/molecular model of the material world. You'll recall, from our discussion of Gay-Lussac and Avogadro, that this model was developed long before chemists could see or measure anything so small as an atom or a molecule. New methods of measurement, such as X-ray and electron diffraction, developed in the early part of the twentieth century, began to confirm the atomic model more directly. Field emission and field ion microscopy, and, more recently, atomic probe microscopes, such as the scanning tunnelling microscope (STM), began to reveal to scientists the geometry of atoms more directly. To deny the atomic/molecular model of matter today is the equivalent of being a member in the Flat Earth Society. In some sense, it is no longer a model but has become reality itself.

All the basic concepts of chemistry, including our understanding of reactions, compounds, the nature of chemical bonds, and the biochemistry of life, rest on the atomic/molecular model. Although you will encounter some variety among textbooks in symbolic representation—Lewis dot structures, match-stick bonds, spherical atoms—do not be misled. There is no disagreement as to the fundamental model of matter; only some variations in the method of representing it. These are, so to speak, models of the model.

The compound propane, for example, can be represented by a number of molecular models (or graphic pictures), as shown in Figure 4.4. Each model is correct and useful in its own way. The *molecular formula* can be easily used for calculations, such as to find the molecular weight of propane or the percentage of hydrogen it contains. The *electron dot formula* (or Lewis structure) shows the manner by which electrons represented by dots form chemical bonds. The *stick formula* can be used to point out which atoms are bonded to each other and to show the nonequivalent locations of the hydrogen atoms. The *dimensional model* and *ball and stick model* allow you to see bond angles and conformations. The *space-filling model* (the hardest one to draw), helps one to predict folding patterns, particularly in large molecules such as DNA and proteins. As new information becomes available, scientists must modify their models to incorporate the new data.

Molecular Formula C_3H_8

Electron Dot Model

```
     H   H   H
     ••  ••  ••
H :  C : C : C : H
     ••  ••  ••
     H   H   H
```

Stick (Valence Bond) Model

```
    H   H   H
    |   |   |
H — C — C — C — H
    |   |   |
    H   H   H
```

Dimensional Model

```
      H   H
 H     \ ▼     H
  \     C     /
   C /     \ C
 /  ▼     /   ▼
H   H   H      H
```

Ball and Stick Model

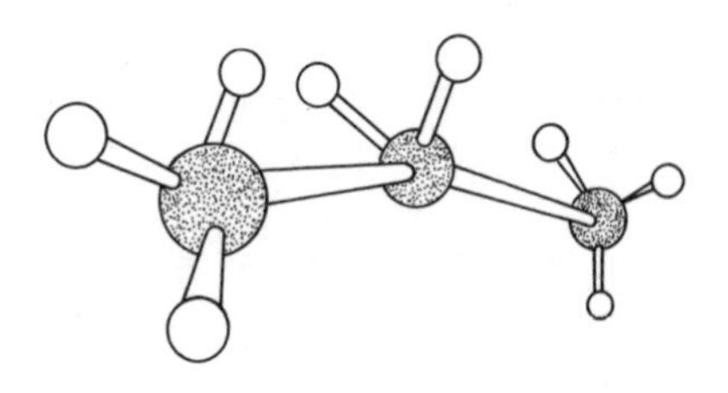

Space-Filling Model

FIGURE 4.4
Propane models.

Modeling a Process

The atomic model in itself does not adequately represent all chemical phenomena. For example, to calculate the amount of heat given off by the dissolving of simple salts in water, chemists construct a kind of model made up of a series of imagined events. They know these events do not actually take place, but the model helps them, as it will help you, to calculate some of the events in solution chemistry. Everyone knows that salt (NaCl) dissolves only after it has been poured into water. But theoretical chemists think of the process as taking place in two steps. First, in the absence of water, the bonds between bind Na^+ and Cl^- are broken by the addition of energy (an *endothermic,* or heat-absorbing reaction). Then, in a second step, they imagine that the now-separated ions, Na^+ and Cl^-, are bound to water molecules (an *exothermic,* or heat-evolving reaction). The amounts of energy involved in both steps one and two can be calculated. By imagining the process in two steps, the model permits a simple subtraction to find the net heat given off.[7]

Do not be surprised by the use of "fictions" in doing science. A famous scientist, Stello, once stated: "Neither mass nor motion is substantially real, only tools of understanding. The path to knowledge is paved with logical fictions."[8]

Models and Theories

In biology a model can describe many things from a hypothesis which is tentative and speculative to a well-established theory. At first, the double-helix structure for DNA proposed by Francis Crick and James Watson was but a speculation. Rosalind Franklin, a British scientist whose data formed the basis of the double-helix breakthrough, had "seen" a vague helical shape in her X-ray

7. The energies absorbed in step one and given off in step two are theoretical calculations, large and of comparable size. Hence the difference between them will be small. In the laboratory, the measurement of heat is difficult but not impossible. What is amazing is that the theoretical results from the model are reasonably close to reality.

8. Quoted in D.J. Kevles, *The Physicists: The History of a Scientific Community in Modern America* (Cambridge, Mass., Harvard University Press, 1971).

diffraction studies of the DNA molecule. But a single helical shape could not account for the number of atoms the researchers believed to be included in DNA, nor for the capacity of the molecule to code so much genetic information. The Crick-Watson model represented the DNA structure as *double* helices—entwining spirals of atoms.

This does not mean that under the microscope DNA looks like the model. Biology students who know the story are disappointed when they don't find a neatly-formed double helix the first time they examine DNA molecules under a microscope and, instead, see what looks like a tangled ball of string, twisted and bent upon itself. It is important to realize that in biology the model doesn't have to look like the actual macromolecule. It only has to be conceptually accurate for it to be useful. The essential features of DNA are the four base molecules (adenine, guanine, cytosine, and thymine); the ways the phosphates connect; and the fact that the hydrogen atoms are located inside, not outside, the helical staircase of the phosphates. Once these essential features of the model were understood, the life sciences were transformed.

When Is a Model Not a Model?

The history of science is full of rules of thumb and numerology that attempt to make sense of an event or a series of observations without a model. One such "rule of thumb" in astronomy is called Bode's Law. It is neither a law, nor was it discovered by Bode. In the statement of this "law," each planet is assigned a number n beginning with Mercury = 1; Venus = 2; earth = 3; Mars = 4; and for no reason except to make the rule work, Jupiter = 6, not 5. You then follow a certain rule (3 × 0, 3 × 1, 3 × 2, 3 × 4, 3 × 8, 3 × 16, 3 × 32 . . .), which gives you 0, 3, 6, 12, 24, 48, 96, ... Add 4 to each number, which gives you 4, 7, 10, 16, 28, 52, 100. . . . and finally divide all these numbers by 10. The final results (see Figure 4.5) are 0.4, 0.7, 1.0, ... These numbers agree strikingly well with the observed average distances of the planets to the sun (measured in astronomical units). But the rule only works if Mercury is counted as 3 × 0, if Jupiter is made 6, not 5, and Neptune is not assigned a number at all.

Planet	Series	Sum/10	Known Distance
Mercury	4 + 0	0.4	0.39
Venus	4 + 3	0.7	0.72
Earth	4 + 6	1.0	1.00
Mars	4 + 12	1.6	1.52
Asteroids	4 + 24	2.8	2.65 (Av.)
Jupiter	4 + 48	5.2	5.20
Saturn	4 + 96	10.0	9.54
Uranus	4 + 192	19.6	19.19
Neptune	—	—	30.07
Pluto	4 + 384	38.8	39.52

FIGURE 4.5

You may wonder about n = 5, which was skipped. The curious thing about Bode's law is that a small "planet" was accidentally discovered in 1801, and its orbital radius was within 1 percent of what was predicted by Bode's law for the missing n = 5. This planet turned out to be the first of what would later be called the asteroids. More than 2,000 asteroids have been discovered in the 190 years since.

There is still no theoretical justification for Bode's Law. Yet the planets Uranus and Pluto, which were not yet discovered in Bode's time, conform to this law if Neptune is excluded. This may not be so arbitrary as it seems if we speculate that Neptune was a latecomer to the solar system, captured by the gravity of the sun. The trouble with this speculation is that the planetary composition of Neptune is typical of the outer planets of the solar system.

Compare the power of Newton's law of universal gravitation as applied to planetary motion to Bode's mysterious "law." After the planet Uranus was accidentally discovered by William Herschel in 1781, astronomers followed its orbit over many years. By 1830 or so, it became clear that Uranus's course deviated from the path calculated by Newton's theory, which included the effects of attraction of Saturn and Jupiter. Some astronomers suggested that perhaps Newton's law of universal gravitation was not really universal and did not apply beyond a certain distance. Others thought that the orbit was being "perturbed" by a gravitational attraction of an as-yet-undiscovered planet beyond the orbit of Uranus. Based on the accumulated observational data on Uranus, two young men, John C. Adams, an undergraduate at Britain's Cambridge University, and J.J. Leverrier in France, independently calculated the position of this Planet *X*. They came up with an exact location for the mysterious planet and, within a single night of search by an astronomer in Berlin, the new planet was sighted at the location predicted. This was how Neptune was discovered in 1846.

The difference between rules of thumb and models is well exemplified in this comparison. A rule of thumb may work for a while, but it has no guaranteed powers. A model, until it is overthrown or modified by contradictory information, is able both to explain long-studied natural phenomena in a new way and in so doing to set a research agenda. Although the concepts of relativity and quantum theory supercede the Newtonian model, no one has ever shown another model to explain the macroscopic world as well as Newton's. One way to think about the "competition" between these models is that the Newtonian world stops at the edge of the atom and near the speed of light. Another way to think about it is that at the limit of a large mass, quantum theory merges with Newtonian mechanics. Likewise, in the realm of "slow" velocities (slow only in comparison to the velocity of light), Einstein's theory of special relativity agrees with Newtonian mechanics as well.

MEASUREMENT

Physics

Science is first and foremost an empirical study of nature; scientists work with observation and measurements. But science is also a human endeavor. After all, it is human beings who are the observers. The relationship between the observer and the thing observed affects physics more than the other sciences because at the level of matter with which physics deals—atoms and elementary particles—the act of measuring one quantity disturbs the measuring of another. Thus, in quantum mechanics, velocity and position of a particle cannot both be measured with precision. This is not because the physicists' instrumentation is faulty, but because the very act of observation itself affects the behavior of the object being observed.

This problem becomes particularly significant in the measurement of subatomic particles. How do you measure the position of an electron? Any microscope requires a light beam shining on the object to be observed. Since light is an electromagnetic wave, its beam will not only illuminate the electron but bounce off it, which will disturb the motion of the electron just as you are trying to locate it. Think about it this way: Imagine you are blindfolded and have to locate a billiard ball on a billiard table. All you can do is to shoot a cue ball in the general direction of the billiard ball you're trying to find. You measure the direction of the returning cue ball. But in the process, your cue ball has moved the billiard ball out of its original position, so you can never know where the billiard ball is *after* the measurement has been taken. It's the same with a light beam and an electron.

This discovery was made by Werner Heisenberg (1901-1976), and is now known as the Heisenberg uncertainty principle. That there would always be a certain degree of uncertainty in measurement was hard for twentieth-century physicists to accept. Theirs had been a stunning four-century progress from confusion to certainty. What the new theory introduced was probability of a statistical nature. Even Einstein never accepted the philosophical implications of quantum mechanics—the fact that there would never be total predictability. To his death, he asserted that uncertainty had no place in nature; it was the "fault" of the model. "God," said Einstein in one of his famous arguments against the quantum mechanical model, "does not play dice."

Chemistry and Biochemistry

Precision measurement was essential to the creation of the atomic/molecular model in chemistry. Chemists realized early that it was very important to determine not just what substances were taking part in chemical reactions, but also in what quantities; first in volumes, then in grams. They did remarkably well, given the limited instrumentation at their disposal: chemical balances, graduated cylinders, and beakers. It was the precise volume ratios of gases at a given temperature and pressure obtained by Guy-Lussac that led to Avogadro's law and eventually to the atomic model, the periodic table, and the field of modern chemistry. Precise measurements of reactants have always been at the heart of discovery in chemistry, even with primitive instrumentation, but chemical measurement now relies on advanced technology.

Naturally, you will not have access to all the tools of high-precision analytical chemistry during your introductory course. The general chemistry laboratory is still largely furnished with the "hardware" of yesterday, although instrumentation is gradually being modernized so that college students can get experience with the newer equipment. At least you will not have to weigh chemicals in the old-fashioned way; college chemistry laboratories now feature automatic, direct-reading balances. Other improvements have come in the means by which chemicals can be differentiated, quantified, and identified.

Different chemicals migrate at different rates in the presence of another medium (a *phase*), an electric field, or gravity. Chemists can use these migrations to separate molecules of one substance from those of another. One of the techniques is called *chromatography*. Molecules of different molecular weights will congregate at different positions in a long tube. Another technique by which such migration can be determined is called electrophoresis (see Chapter 3). Here an electric field is created and different molecules migrate at different rates. Once molecules are separated by their characteristics, the next task is to determine their molecular masses. This is done with the aid of *mass spectrometers* of various kinds. Instrumentation is also available to locate atoms in their molecular environment. Chemistry research laboratories are equipped with *nuclear magnetic resonance* (*NMR*) spectrometers and *electron paramagnetic resonance* (*EPR*) spectrometers which help chemists identify and locate atoms and molecules.

Optical instruments are useful in chemistry because specific atoms and molecules absorb and emit light of certain wavelengths. The wavelengths are their signatures; each element has its own. The element helium (see Chapter 3) was discovered when chemist Joseph Fraunhofer noticed a series of black lines in the otherwise rainbowlike spectrum of the sun, absorption lines that did not correspond to the signature of any known atom. (Atoms cannot bond chemically in an environment as hot as the sun; hence there are no molecules.[9]) His conclusion was that the new signature belonged to an element not previously observed on earth and he named it helium. Today, chemists combine lasers, fiber optics, and computers to make powerful tools for measuring and identifying chemical elements and compounds.

Spectroscopy covers the entire electromagnetic spectrum, well beyond our vision, from high-energy gamma rays to low-energy radio waves (Figure 4.6). This permits scientists to study natural events from the formation of molecules to the Big Bang.

As Figure 4.6 shows, the visible range is a very narrow portion of the electromagnetic spectrum, yet it is sufficient to provide vision to the organisms that evolved on this earth. What is more remarkable is that the water vapor in the earth's atmosphere absorbs almost all the electromagnetic radiation between the infrared and the ultraviolet except for the very narrow visible range. This makes it possible for us to see, and more important, to contemplate, the planets and the stars. If ancient humans had not been able to speculate about the heavenly bodies, who knows what questions in science and in religion might have gone unasked?

NONSCIENTIFIC APPLICATIONS

Mathematics, measurement, and models are not unique to science. In recent decades and partly because of the success of the natural sciences, the social sciences have become more empirical, more quantitative, more model-driven, in short, more "scientific." Thus, the study of natural science paves the way for the study of other disciplines. Economics, particularly microeonomics, relies heavily on the concepts of calculus. And if you know the techniques of calculus, you will be way ahead. Business, especially decision-making in business has become an empirical science, resting on the mathematical model of "game theory," among other principles, such as linear programming. Debates among psychologists—between "Skinnerians," for example, who believe behavior can be understood as a series of "neural events," and "func-

9. There are no neutral atoms either. In the high temperature of the sun, all atoms are ionized, i.e., charged; such collections of ions are called *plasmas*.

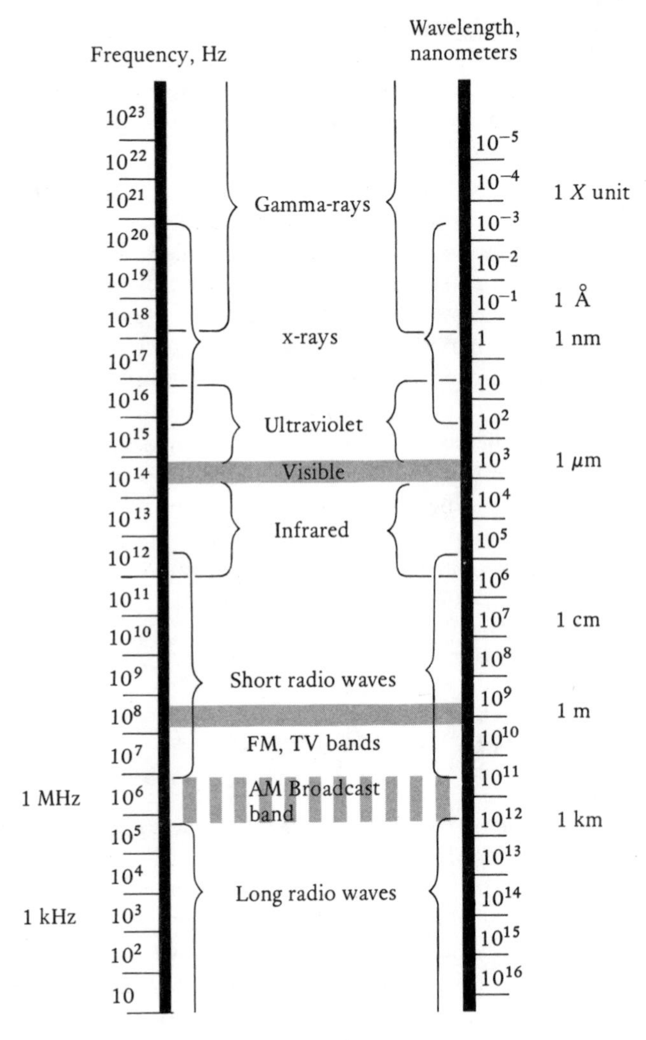

FIGURE 4.6

tionalists" or "dynamic psychologists," who believe the organism itself plays a role in its response—are debates about models. Indeed, the latest model of the brain provided by neuroscience is being "imported" back into physics and computer science as the model of *neural networks* applied to the construction of the hardware and software that will produce a computer that can "learn."

Psychologists, sociologists, and political scientists don't believe that individual actions or events can be predicted. Humans are more than "matter"; motivation involves more than "forces." But they do believe that when humans act collectively, i.e., as ethnic groups, or nations, or in crowds, their behavior follows certain patterns or "laws." Hence, your coming to terms with "laws" and "models" in natural science will serve you very well in just about any profession you might choose.

5

What's Ahead in Science?

On February 26, 1991, Linus Pauling, who has won two Nobel Prizes (one for chemistry, one for peace), celebrated his ninetieth birthday in Pasadena, California. His colleagues from the California Institute of Technology, where Pauling taught and did research from 1922 until 1963, joined with colleagues in chemistry from around the world to wish him well in a way that scientists prefer to a mere social celebration. They staged a daylong symposium in his honor on "the future of chemistry." The event would have been just as dramatic had the chemists focused, instead, on what has happened in chemistry and its related sciences during Pauling's lifetime. In the span of that one life, 1901 to 1991, our understanding of nearly everything that was believed known about nature has changed; Pauling and his generation of scientists are responsible for several of those scientific revolutions.

Think of the science Pauling learned as a young person compared to the science his generation is leaving behind. About 1914 when Pauling was beginning high school, the periodic table of the elements still had unfilled boxes representing elements chemists expected to find but had not yet discovered or named. Moreover, the periodicity of the elements on which the table was based was still a mystery. The electron itself had only been discovered fairly recently (1889). The way in which chemical properties depend on electron configuration was not yet understood. Chemistry, it was thought, was generally unrelated to physics, and neither was particularly pertinent to biology, which in Pauling's adolescent years was still a "field science"—divided between botany (the study of plants) and zoology (the study of animals). The principle and even the patterns of inheritance were known (since the time of Gregor Mendel), but not the mechanism for transmitting traits. No one had the faintest idea that the coding for genetic information resided in the structure of one nucleic acid (DNA). In physics, the nuclear model of the atom was just emerging. The Curies had earlier isolated radium and discovered the impermanence of certain elements. But even the Curies could not have known that their studies of radioactivity would lead, in less than half a century, to an understanding of chemical bonding (Pauling's own work), and to the manufacturing of a dreadful family of weapons that Pauling would never cease to oppose.

Neither Pauling's teachers nor Pauling himself could have predicted all of what was to come in just one scientific lifetime, so it would be foolhardy to try to predict all that may happen in yours. But some trends can be seen, just as the discovery of the nuclear model of atoms paved the way for an understanding of the periodic table. To get some sense of current trends in scientific research, we asked a number of working scientists to tell us, in their own words, "what's ahead" in their areas of science; where they think progress is likely to occur. What follows, then, is not a comprehensive survey, but rather a set of personal views.

CHEMISTRY

"Your question, 'what's ahead in science?' made me realize how uncomfortable scientists feel when asked to predict the future," writes Joan Valentine, professor of chemistry at UCLA. "I think the reason is that many of the most exciting discoveries are of phenomena that we wouldn't have dreamed of 10 years ago. Of course, this is what makes science exciting, but it also makes us feel a bit foolish. Scientists don't like to admit it when they are surprised." Valentine didn't start out to be a chemist. She says her childhood heroes were Sherlock Holmes and Nancy Drew. But after an undergraduate experience in a research lab, scientific research seemed to her to be a "safer way to be a detective than becoming a private investigator," and so she went on to earn a Ph.D. in inorganic chemistry.

Today, she heads a research group of 12 postdoctoral associates, graduate students, and undergraduates studying the properties of biological systems that contain metal atoms. It is interesting that although she is an inorganic chemist, Valentine is at work on biological systems. The reason is that although the bulk of the molecules in our bodies are organic (involving mostly carbon), certain critical biochemical molecules involve metallic (inorganic) elements as well; for example, iron, copper, and zinc. Hence, some chemists in biochemical and biomedical research are called bioinorganic chemists. Valentine's group, in fact, is studying both the enzymes that catalyze reactions of oxygen from the air and the enzymes that guard against some of the toxic effects of intermediates produced by oxygen metabolism. That's why their work is so relevant to medicine.

As to what's ahead in chemistry more generally, Valentine points to the development of new types of materials with unusual properties. "We pretty much understand what holds atoms together in molecules and what holds molecules or atoms together in bulk solids," she says. "But recent research on small clusters of atoms has uncovered the fact that such small clusters have unique properties different from isolated atoms and also different from the properties of the atoms in bulk materials. If chemists can understand the rules that govern such small clusters, they will have a more complete understanding of what holds matter together."

Another important area of research, says Valentine, is surface science. "A bulk material has different properties in the middle, where all atoms or molecules have neighbors in all three dimensions, than it does on the surface where the atoms or molecules are exposed. Understanding the nature of the bonding and interactions of such surfaces is very important because it is these properties that allow us to bind two kinds of materials together (such as sticky tape and paper)."

Measuring the Molecular Event

Perhaps the most exciting paper delivered at the Pauling symposium had to do with the fact that chemists now, for the first time, are able to monitor and measure a molecular event. Ahmed Zewail discussed femtochemistry, a new field that he and the late Richard Bernstein created and named; it describes how, using lasers, chemists can now excite molecules and track the movement of electrons and atoms. Using light pulses of femtoseconds duration (even shorter than the time it takes molecules and atoms to react), they are now able to "watch" atoms rattle around and electrons jump between them. As Harry Gray, professor of chemistry at California Institute of Technology and a participant in the Pauling symposium explained, "Lasers allow us to examine chemical reactions that chemists have never been able to monitor before, reactions that take only trillionths or even fractions of trillionths of seconds."[1]

What happens in femtochemistry is this: Very short light pulses, some of them operating at a fraction of a trillionth of a second (femto- means 10^{-15}, or one thousandth of a trillion; hence femtosecond),[2] are beamed at molecules. The excited molecules absorb light differently from the way they do in their unexcited state, so a second monitoring beam can follow the changes going on—the very movements of atoms as they separate from molecules. (This is analogous to using a strobe light to analyze the motion of an athlete.) Lasers have been available for the past two decades. What is new, Gray tells us, is that the pulses can now be even shorter in duration than the molecular events themselves. Hence, chemists can follow—in a sense "photograph"—primary chemical events. As Gray sees it, chemists' growing ability to measure molecular events is going to reshape their fundamental understanding of the reactions of molecules and atoms.

Molecular Synthesis

Another expanding area of chemistry is the synthesis of whole new classes of molecules, some of which will be used in medical applications, to prevent environmental degradation (such as finding substitutes for CFCs), or for environmental cleanup. These new molecules will depend on, and contribute to, a better understanding of, chemical processes. Valentine says that tools are already available to make synthetic molecules from biological materials, and then to direct the synthetic capabilities of living cells

1. See Harry Gray, Priestley Medal Lecture, *Chemistry and Engineering News* (April 15, 1991), p. 17.
2. Other commonly used prefixes are discussed in Chapter 2.

to synthesize these molecules themselves. In this way, the immune system may be "tricked" into making antibodies that duplicate essential features of enzymes, so that they can convert one type of molecule into another.

Artificial enzymes, together with new biomaterials will help scientists conquer the problem of *biocompatibility* in designing and manufacturing artificial organs, implants, and prosthetic devices, for example. Valentine describes the work in this field as an important new basic science. There was a time when scientists would have considered these to be applications, but as the line between basic research and applications becomes harder and harder to draw, these areas are attracting pure scientists as well. Valentine says they are "intellectually stimulating" and, because their uses are so important, "personally gratifying" as well.

New Pharmaceuticals

One of the many fascinations in the pure science of molecular chemistry is that some molecules are identical to other molecules except in their geometrical orientation: they are mirror images of each other. Chemists call this difference "handedness." Just as right would appear as left in a mirror, so some organic chemicals appear in nature in "l" forms (left-handed) and others in "d" forms (right-handed). They are both the same structure, made up of the same atoms but oppositely spiraled—one being the mirror image of the other.[3]

Enzymes catalyze chemical transformations if they "recognize" the substrate (the substance that undergoes the chemical transformation). An enzyme that recognizes a molecule in one three-dimensional form won't recognize its mirror image, just as a right-handed glove will fit the right hand but not the left. Although a few organic substances are found in both "l" and "d" configurations in nature, by far the bulk of natural biochemicals come in one configuration or the other. Today, chemists can synthesize molecules that have a specific handedness and, by doing

3. A light beam going through a "d" form (called "d" after the Latin *dexter* meaning "on the right") will rotate a plane one way; a beam going through an "l" form will rotate the plane the other way. Molecules that include both "d" and "l" forms won't show this effect.

so, regulate the physiology of a substance. "What this means for drug design," says Michael Doyle, professor of chemistry at Trinity University in San Antonio, "is that by constructing molecules of a single handedness and not their mirror image, chemists will be able to prepare pharmaceuticals that enhance the disease-fighting capability of certain drugs while removing negative side effects that are due to the mirror-image molecules." Doyle himself has patented a family of new catalysts which can selectively produce chemicals with a particular handedness. Remember, catalysts are substances that assist chemical reactions without being changed by them. Doyle's catalysts are of immense value in making pharmaceuticals and agricultural products more efficient.[4]

MICROBIOLOGY

The development of new biological materials is an interdisciplinary effort. Materials science is no longer separate from biological science, says Doyle. John Spizizen, a microbiologist, agrees.

Spizizen is a professor emeritus in microbiology at the University of Arizona. He has retired from his position as chair of the department of microbiology in the school of medicine, but is still active in research on genetically inherited disease. The field in which Spizizen spent most of his professional life, microbiology, was not yet a field when he was a student, so he trained himself in biology and biochemistry. Bacteria and viruses were known to scientists, but the instrumentation to study them in any detail was not yet available. Scientists who studied microorganisms were called bacteriologists, and were associated with medicine or agricultural research. Today, microbiology includes the study of all organisms that are microscopic, or even submicroscopic, in size, including bacteria, viruses, and fungi.

From the beginning, Spizizen decided to specialize in viruses, both because of their significance in human disease and because of their peculiar parasitic quality. His interest was in a particular class of virus, called *bacteriophage*, each of which has its own unique bacterial host. "What makes viruses interesting," Spizizen explains, "is that they can survive for long periods in an inert

4. *Technology Forecasts*, Vol. 23, No. 6 (June 1991), p. 1.

state. But, in order to metabolize food and to reproduce, they have to invade some other living cell. They are the ultimate parasite, an intracellular one." When viruses invade a host cell, they can sometimes employ the host's own DNA to reproduce. Eventually (within 20 minutes in most cases), they and their progeny burst the cell, leaving a hole on the host cell that is visible under a microscope. This is what makes it possible to study viruses experimentally. Although they are too small and short-lived to be counted individually, their numbers can be determined experimentally in the burst bacteria they leave behind. "What made viruses interesting for scientists," says Spizizen, "was this discovery that they could be counted, however indirectly, in a laboratory setting."

The science of microbiology has become important for several reasons. Microorganisms are easier to produce and to study than are whole animals, or even single (complex) cells of larger organisms; small as they are, they are complete or nearly complete organisms. So biologists can learn a great deal about the mechanisms of life from them. In addition, their extremely short life cycles—viruses and bacteria reproduce themselves in minutes—and the fact that their progeny can be counted makes it possible for geneticists to collect multigenerational data in the laboratory. Also, there is an enormous variety of microorganisms. Spizizen tells us that new microorganisms are discovered every day. Some of them are simply new to us, because we didn't know they existed; some are actually newly formed, because bacteria and viruses are mutating all the time. The discovery of a living organism no one has ever identified before can be a very exciting experience for anyone in science.

New Microorganisms

Sometimes the discovery of a new species of microorganism is born in tragedy. About 15 years ago, some members of the American Legion attending a convention in Philadelphia were stricken with a severe, and in many cases fatal bacterial infection. The bacterium that killed them had never been seen before; when it was finally isolated, scientists named it *legionella,* after its victims. The same is true of the AIDS virus, which devastates the human immune system; microbiologists did not encounter it before the

1980s. New forms of bacteria and viruses are always being created in nature. Hence, scientists have a great deal to do just to keep up with these new organisms.

But nature is not the only source of new microorganisms. In very recent decades, as gene sequencing has become better understood, scientists are learning to "engineer" new microorganisms. By injecting certain genes into existing organisms, we can alter their biochemistry, causing them to produce certain enzymes and other proteins that we humans need. As a result of this new technology, there is a wide open future for applied microbiology. Microorganisms can be engineered, for example, to consume certain waste substances, such as plastics, that are otherwise costly or even impossible for industrial civilizations to eliminate. When these microorganisms already exist in nature, they can be "harvested"—farmed like crops—in huge vats, stored, and even deep frozen, to be thawed and used later, to consume the remnants of an oil spill. They can also be destroyed at will, simply by depriving them of food.

A new bacterium, developed by a scientist or engineer for a particular industrial process, can be patented, just like a new chemical or any other kind of industrial device. The issue of patenting living matter was sufficiently new and controversial to land in the courts, where it was recently ruled that a living organism is a patentable entity.

"Bacteria are very flexible," Spizizen says. "Scientists can not only alter their genes but they can affect their production of, say, a particular enzyme, one that will attack a particular waste product, for example, by altering the environmental conditions, temperature, amount of food, etc., in which the bacteria are made to grow." All of this is intended to get them to do the job they are being harvested to do. If a particular bacterial enzyme is desired, scientists can arrange to have that enzyme amplified in the particular bacterium. Spizizen calls these "genetic tricks" that can be played on the organism.

Even though viruses are becoming better understood, Spizizen doesn't think viruses will ever be as useful as bacteria in manufacturing chemicals or in eliminating wastes. That's because, as already noted, viruses cannot reproduce without a host cell. However, insofar as they can alter the biochemistry and physiology of a bacterium, they, too, might have applied industrial uses in the future. For example, there is a bacteriophage (a virus) that makes the diphtheria bacillus virulent. If the bacteriophage is not present, the diphtheria bacillus doesn't cause the disease. If one kind of bacteriophage can do this, it might be possible to use another to render a virulent bacterium harmless.

Microorganisms have always been known to create certain desired products; sauerkraut, yogurt, green olives, pickles, soy sauce, the leavening of baked goods, and the fermentation of wine are all the results of bacterial action. In the 1940s, when penicillin was extracted from mold, an era of antibiotic manufacture employing bacteria and fungi was ushered in. Insulin, for example, used to be obtained from the pancreases of cows. Today, with recombinant technology (genetic engineering), the genes that make insulin in the cow can be removed from the cow, injected into E. coli (bacteria), and produced for human use in massive quantities. This technology is not limited to insulin production; there are many other possible applications.

PHYSICS

John Rigden is a physicist, biographer (of the physicist, I.I. Rabi), former editor of *The American Journal of Physics*, and director of programs at the American Institute of Physics. So he not only does physics, but thinks long and hard about what makes physics uniquely different from the other sciences, and how this uniqueness will affect the future of physics in the next few decades. He notes that unlike chemistry and biology, which have always benefited from developments in physics (and continue to do so), physics can look to no other science for direct assistance. It is entirely dependent on itself. Where, then, will any "new physics" come from?

Rigden believes that physics is still building on two great revolutions that occurred earlier in this century: relativity and quantum mechanics. It is only since the 1930s that physicists have understood that, in addition to gravity and electromagnetic forces, there are two other forces in nature: the so-called "strong force," which binds protons and neutrons together in a nucleus; and the so-called "weak force," which comes into play when certain nuclei decay. Einstein's aim was to unify all these forces in a single theory. In 1974, that aim was partly realized when physicists unified two of the forces, the electromagnetic and the weak force, and the standard model emerged to unify all but the gravitational force. What remains is the challenge of completely integrating the three forces and then incorporating gravity as well. This is what Rigden means when he writes, "So rich are the implications of quantum mechanics for investigating the intricacies of matter that, as we approach the next century, there is still much work for physicists to do."

Take the proton, for example. In the previous models, the proton was presumed to be stable. But as the existence of more and more elementary particles (such as quarks) is confirmed, questions are being raised about the stability of the proton (protons are made up of three quarks). Is the proton stable or does it eventually decay? There are other unanswered questions about quarks, such as whether the so-called "top quark" exists. The unification of forces also raises questions about the dynamics of the Big Bang—what may have happened in the first microseconds after the universe began, when the unified force was presumably still unified.

At the same time, questions are raised by the fact that matter is not uniformly distributed throughout the universe as one would expect from the Big Bang theory. Astronomers see patches in the sky where even the most powerful telescopes don't reveal as many galaxies as in other parts. Is there a problem with the model? Will new experimental techniques reveal still more troubling discrepancies? Perhaps the principles on which the theory of the origin of the universe is based may need to be reconsidered.

New eras in physics have often been ushered in by new instrumentation, Rigden reminds us. The *vacuum pump*, invented in 1885 by Heinrich Geissler, led directly to the discovery of the electron. The *vacuum triode*, invented in 1904 by Lee De Forest, opened the way to the design of new electronic circuits with which physicists could detect and amplify signals, making measurements more sensitive and more precise. The *transistor*, invented by John Bardeen and others in 1948, replaced the vacuum tube and made it technically possible to pack tens of thousands of these tiny transistors into small spaces, extending the power of electronics still further. The *laser principle*, known by 1955, made it possible for physicists and chemists to use coherent sources of light to tease atoms and molecules into revealing their most subtle properties;

without the laser, there would be no femtochemistry. Although the computer is already widely used, its full impact on scientific research and on our lives is still ahead.

"Physics," says Rigden, "depends on ideas from within and devices from without. So long as there is a human species, intrigued by the physical universe, they will probe what is not seen to better understand what is seen. And when they do this, they will be doing physics."

Physics of Complex Systems

One of the implications of classical physics is that once a system gets going its behavior is predictable (sometimes called *deterministic*), and can be described in a mathematically exact way. Moreover, most of these systems, the ones studied so far by physicists, are linear. This means that they respond proportionately to any change in their input. Double the applied force and the resulting acceleration will be doubled. Halve it and the resulting acceleration will be halved. Until recently, physicists did not like to deal with the dynamics of *nonlinear* systems.

But much of nature is *nonlinear.* Weather, turbulence in air flow, the path of lightning, and your heartbeat are but a few examples. The way physicists handled the dynamics of these systems in the past was by approximation; that is, the nonlinear segment of the phenomenon was ignored. Today, the study of nonlinearity is one of the most exciting topics in physics and biology. Aided by powerful computers, scientists can study phenomena that defied them before.

One feature of the differential equation (see Chapter 4) is that before you can describe the details of the future behavior of any system, you need to know, with some precision, what it was like at the beginning. Scientists call this the "initial condition." Take a motion problem. If you know the speed and position of an object at time zero (its initial condition), you can use the differential equation for that system (Newton's second law of motion) to describe its speed and position at every instant thereafter. This is the power of classical physics and why the mathematics of differential equations is vital for physical science. An important characteristic of these linear systems—one that you might not

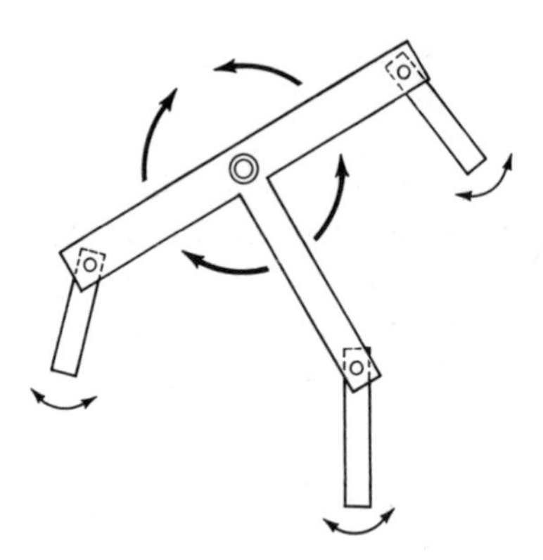

FIGURE 5.1
Three-armed pendulum.

notice as you first encounter these problems—is that a slight change in the initial condition will only slightly change the outcome. This becomes important later on.

A swinging bar pendulum (when the amplitude is small) is a typical linear system. If you give the pendulum a bit more push at the beginning, it will end up swinging in an only slightly wider arc. But if the pendulum is three-armed, and each arm has a freely swinging dangling bar (see Figure 5.1), the system is nonlinear, and a small change in the initial condition will have a large and unpredictable effect.

In the pendulum illustrated in Figure 5.1 (which is derived from an exhibit at the San Francisco Exploratorium, a museum of science), the effect of the dangling bars is that every time you start the pendulum, even if you try to do it exactly the same way with exactly the same amount of force, the swing will vary uncontrollably. This is typical of a nonlinear system: a very small alteration in the initial condition results in an unpredictable departure from the previous pattern. Because these systems are irregular in interesting ways, physicists and mathematicians refer to them (and not as a joke) as "chaotic."

Chaos

Some physicists and mathematicians believe that *chaos* is the century's third great revolution in the physical sciences, after relativity and quantum mechanics. One journalist, James Gleick, explains the concept this way:[5]

> Watch two bits of foam flowing side by side at the bottom of a waterfall. What can you guess about how close they were at the top? Nothing. As far as standard physics was concerned, God might just as well have taken all those water molecules and shaken them personally.

Traditionally, says Gleick, when physicists saw complex results, they looked for complex causes. When they saw randomness, they thought they would have to build in some "noise" or "error" into their theory. The modern study of chaos began, he writes, with the "creeping realization in the 1980s that quite simple mathematical equations could model systems every bit as violent as a waterfall. Tiny differences in input could quickly become overwhelming differences in output. . . . Only a new kind of science could begin to cross the great gulf between knowledge of what one thing does—one water molecule, one cell of heart tissue, one neuron—and what millions of them do."[6]

As physicists use the term chaos, they don't mean that chaotic systems are incapable of being predicted. Rather, they mean that such systems are extremely sensitive to change in initial conditions. See how this shows up graphically. Figure 5.2 depicts a linear system. Note how slight changes in the starting point shift the resulting curves only slightly. Note also that one could interpolate (determine from the other curves) the shape of another curve between any two curves. Figure 5.3 shows a nonlinear (chaotic) system at work. Note how each slight change in the initial conditions changes the shape and slope of the resulting curve altogether, and that there is no way to interpolate the curve produced by other initial conditions.

Weather is a good example of a nonlinear system. Everyone knows that long-term forecasting is extremely uncertain, although short-term forecasting is more reliable. This is because the flow of air masses and the heat exchange among these masses and with the earth behave nonlinearly, giving rise to complex outcomes.

5. James Gleick, *Chaos* (New York: Viking, 1987), p. 8.
6. Ibid.

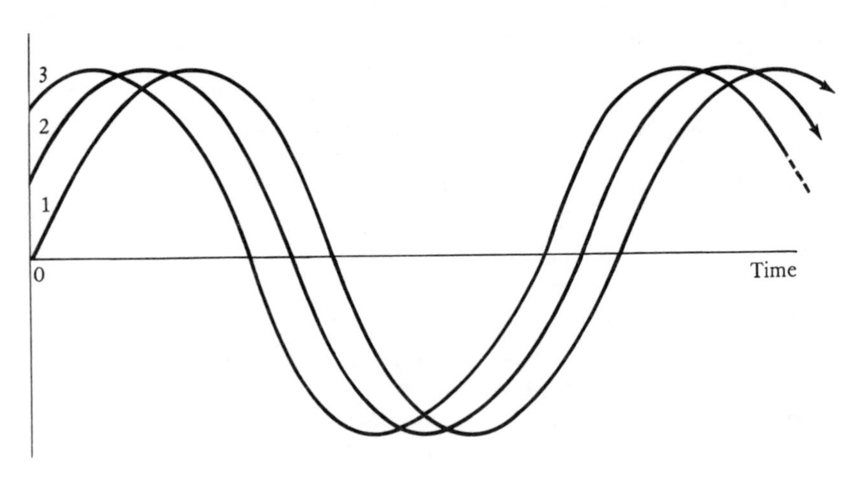

FIGURE 5.2
Angular position of a linear pendulum.

Because change in initial conditions is so critical to weather patterns, weather can be considered a "chaotic" phenomenon. Insiders familiar with chaos theory joke about "the butterfly effect" in long-term weather forecasting, meaning that the flutter of a butterfly's wings in Beijing could cause storm conditions a month later in New York.[7] Although no one can prove that a particular butterfly caused a particular storm, the image is a good one, for it captures the essential feature of nonlinear systems; a small change in the initial condition can have a very large outcome.

PLANETARY SCIENCE

Observation of the heavens is where science began. It was the fact that starlight can be perceived by human vision that piqued humankind's curiosity thousands, perhaps tens of thousands, of years ago. At first, the regularity of the apparent movement of the sun around the earth and of the changing seasons reassured our ancestors and often led to belief in a deity that controlled these events. Later, people realized that this regularity was due to me-

7. Ibid, p. 9.

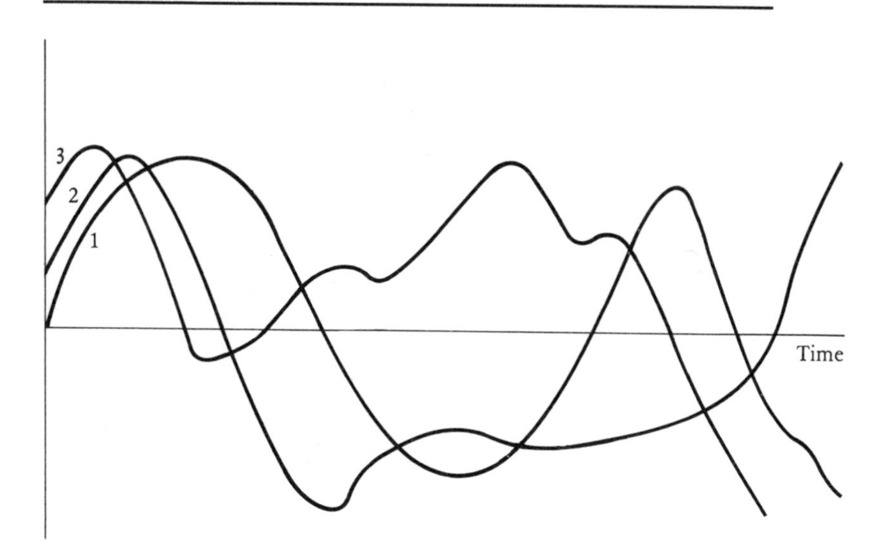

FIGURE 5.3
Angular position of the middle bar of a nonlinear pendulum.

chanical forces that could be studied and understood. Today, with the availability of ever more sophisticated telescopes—some of them orbiting the earth—and actual space travel of vehicles carrying scientific instruments, our knowledge of the planets, the stars, and the galaxies is hundreds of times greater than it was in the past, but our curiosity and our awe are no less than those of the Ancients.

So great is the interest in these matters that what used to be called astronomy now has developed into two fields that are in some ways moving apart from each other. One is stellar and galactic astronomy, and the other is the study of the planets, now called planetary science. To find out what's ahead in planetary science, we contacted Timothy Swindle, a professor in that field at the University of Arizona in Tucson. In reply to our query, he wrote:

> Although planets and meteorites have little impact on our everyday lives, thinking of the Earth as just one of a bunch of planets can actually give us insight into its history and even its future. For example, geological and atmospheric processes, such as erosion and plate tectonics (continental drift), have

erased most of the clues to the Earth's formation, but by studying other places that have been less active in the recent past, such as the Moon, we can piece together what must have happened in the distant past. Also, we can learn about ways in which atmospheres can or cannot evolve by studying those of Mars and Venus, which probably were similar to Earth's atmosphere when the three planets formed, but later diverged into wildly different (though equally inhospitable) places.

Note that he refers to Mars and Venus as "places." For the past 30 years, Swindle explains, spacecraft have turned astronomical objects into places that can be explored. The "first-look" phase is largely complete, but these places are still largely unknown, and much of the information from the first round of exploration isn't fully understood, so there is more work still to be done. Another source of new information is the study of extraterrestrial material—virtually a nonexistent field 30 years ago—although a few people have always studied meteorites. Today, scientists are poring over moon rocks, and, as better and better equipment makes it possible to look at smaller and smaller pieces of extraterrestrial material, learning more about conditions in the early solar system.

Interplanetary Exploration

What's ahead, depending on the amount of government funding, are more distant interplanetary missions like the Viking mission to Mars and the Voyager flyby of Jupiter. The primary targets for the next couple of decades, Swindle believes, will be Mars and the moon, even if the current hope of lunar bases and human expeditions to Mars is not fulfilled. For the moon, there are plenty of follow-up questions that have been produced by 20 years of analyses of lunar samples. Many of these questions involve the detailed geology of specific places, but there are some intriguing broad concerns to be addressed as well. For instance, since the last dedicated lunar mission nearly 20 years ago, a new idea—the collision of a planet-sized interloper with the earth—has become the leading theory for the origin of the moon. Like any good scientific theory, it makes testable predictions, but many of them cannot be tested until spacecraft return to the moon. For example, the size (or even the existence) of a metallic lunar core like the earth's core is crucial to the theory. The Apollo missions did not detect such a core, but there may be one large enough to have important effects on lunar chemistry. There are also ongoing debates about the history of impacts on the moon, some of them tied to the question of whether impacts of large objects come in bursts, and if so, why.

Mars is the planet most similar to earth. But how similar is it? In particular, is it similar enough for life to have formed? There are channels on the surface of Mars that may have been cut by liquid water. But there is no consensus on exactly when they were cut, for how long liquid water might have been flowing on Mars, whether there were standing bodies of water (lakes or oceans), why or when conditions changed, or how much water might be present as ice today. All of these can be addressed, if not completely answered, by further spacecraft journeys. And these answers will lead to an even more intriguing question: If there once was standing water on Mars, conditions might have been much like those thought to have been present on earth when life formed. Did life also exist on Mars? Although the Viking missions detected no evidence of life at present, the issue is likely to remain controversial for decades to come.

Earth-Based Technology

Although interplanetary missions virtually guarantee dramatic, usually photogenic, increases in our knowledge, much excitement is still being generated by earth-based technology. For example, our increasing computing ability makes it possible to construct more and more realistic models of complicated systems, such as the *solar nebula* (the cloud of gas and dust that gave birth to the sun and planets), colliding planets (which might have produced the earth's moon), and planetary atmospheres (including earth's presently evolving one). Static high-pressure experiments can now almost simulate pressures at the core of the earth, making many theories of the earth's interior testable. Sophisticated telescopes, both on earth and in low-earth orbit, may make it possible to detect planets around stars other than the sun. Although theories of the formation of the solar system predict that such planetary systems should be common, we know of only one, our own. A number of techniques have suggested the existence of planets that orbit around specific nearby stars. One of these, says Swindle, is likely to bring solid proof of another planetary system in the next 20 years.

Planetary Clues to Earth's Past

From studies of the composition of meteorites, scientists now speculate that it was a collision with an asteroid some 65 million years ago that probably caused the extinction of the dinosaurs. The fact that there is a stratum in earth rock with unusually high concentrations of the element iridium (an abundant element in the composition of meteorites) suggests a collision at just about that time. A colliding asteroid might have disturbed the environment sufficiently to cause the extinction of all but the rat-sized mammals from which humans descended. One theory is that the dust and debris from the collision caused a kind of nuclear winter, blocking out the sun's rays for months; another is that temporary heating of the whole atmosphere killed certain species, and giant coastal waves (tsunamis) drowned others. A third theory is that the chemical effects of an extremely acidic rain killed off the large mammals and their food supply. In the discussion of these theories, planetary scientists contribute a knowledge of meteorite chemistry, and geologists contribute an ability to date rock strata. Thus are geology and planetary science linked.

GEOLOGY

Geology cannot be studied in isolation; it depends on chemistry, physics, biology, and mathematics. Some geologists specialize in its chemical aspects. They study the elements in the earth's crust and the chemical reactions that cause materials to form out of those elements. Others are more interested in the forces that contribute to the changes in geological formations. *Mineralogy*, the study of minerals that constitute the rocks of all planets (not just our own) depends on inorganic chemistry. *Paleontology*, the study of the evolution of organisms, is biologically based. Mathematics is useful, too, since geologists employ computer modeling to understand the internal movement of the earth's layers. Thus, for students with wide-ranging interests in science and a desire to work both in the laboratory and in the field, geology is an exciting area of research.

We asked Jill Schneiderman, a working geologist and a professor of geology at Pomona College in Claremont, California, to help us convey to our readers what's happening and what's ahead in geology. Schneiderman became interested in geology very early in her education. "Having learned about the duration of geologic time and the immense age of the earth as an adolescent, just at a time when I was both awed by the spectacle of some of the earth's natural wonders and distraught by the ways in which I saw humans treating the planet, I became convinced that the only way to save the planet was to understand how it works." At 16, she was able to take part in a summer environmental science program for high school students, which intensified her interest. After graduating from Yale University with a major in geology, she worked first for an environmental "think tank." Then "itching for adventure and some real geologic fieldwork," she went to work for Anaconda, making geologic maps of western Alaska by means of helicopter reconnaissance. In graduate school, at Harvard University, she specialized in *metamorphic petrology*, the study of rocks that occur in collisional mountain chains such as the Himalayas, the Alps, and the Appalachians. Her studies have taken her thus far (and she is only in her early thirties) to Pakistan, Finland, Kenya, Alaska, and much of North America, where she not only observes rocks, but enjoys getting to know other people and cultures.

During the summer of 1991, Schneiderman and a colleague led a group of undergraduate students to the then Soviet Republic of Georgia (now simply Georgia) in order to study the geology of the Caucasus Mountains. Her students joined with Georgian students in this endeavor. What makes the Caucasus Mountains particularly interesting is that they are located between the Himalayas to the east and the Alps to the west. All three mountain chains were formed by the same type of geologic process. That is, the collision of two plates of the earth's crust pushed up these high mountains. In addition to the geology of Georgia, her students were also introduced to its people and culture. When Schneiderman's group returned home, they were accompanied by a delegation of Georgian students and faculty who were then shown the geology of portions of the United States.

Currently, Schneiderman's research is on the San Gabriel Mountains in southern California. The San Gabriels, like many of the mountains of the western United States, consist of five blocks of suspect *terranes* (geological formations); they are called "suspect" because geologists don't yet know where they came from. With funding from the Petroleum Research Fund of the American Chemical Society, Schneiderman will analyze the chemistry of minerals in rocks from two of these blocks to try to determine their geologic history.

Environmental Problems

Looking ahead, Schneiderman predicts that students selecting geology today will have many opportunities to help solve the world's environmental problems. One set of problems has to do with global warming, another with the recently documented hole in the ozone layer. Two areas of geology are of particular relevance to such environmental issues. The study of the dynamics of fluids helps explain atmospheric changes. The study of the movement of ground water is extremely useful in water supply and water contamination. Schneiderman says:

> We dispose of the massive amount of hazardous wastes that humans produce by burying them in the ground. But water percolates through the soil and rock in which this waste is buried, thus contaminating our water supplies. In the future, hydrologists and geologists will help society plan ahead, perhaps in innovative ways, so that the disposal of waste does not poison our water supplies.

If, in fact, we do enter a period of global warming, diminishing water supplies will become a major problem, especially in arid portions of the world such as the western United States and Africa. In that case, *hydrology* (the study of the earth's water) will be of immense importance in solving drought-related problems. One strategy will be to search for new water sources, perhaps by transporting icebergs from polar regions; another will be to desalinize sea water.

Geologists deal not only with long-term changes, but also with immediate natural catastrophes, such as earthquakes, volcanic eruptions, and tsunamis (earthquake-generated gigantic ocean waves). To design structures that can withstand earthquakes, construction engineers call upon geologists for help. And when prevention fails, geologists are involved in relief planning for such natural disasters. In the past, trained geologists were employed in the energy industry, largely in exploring for new supplies of oil, minerals, and natural gas. But as the world's dependency on nonrenewable energy sources wanes, they will be even more valued in researching and developing renewable energy sources. *Geothermal energy* (produced when the earth emits heat, as in the Salton Sea in California or the geysers in Yellowstone, Wyoming) is a potential replacement for coal, oil, and gas. Tidal energy is a potential replacement for nuclear and hydroelectric power plants. For all these reasons, Schneiderman believes that majoring in geology with an environmental studies concentration is a wise choice for students who like science and who, like herself, care deeply about what environmentalists call "spaceship earth."

Social Responsibility of Science

With all the science that is left to do, there should be something for *you* to do in science. Although we celebrate the enormous achievements in the various fields of science and look forward to even more in the decades ahead, we cannot end this book without a cautionary note. It is important to realize the distinction between pure science—the search for knowledge about nature—and the uses to which human beings put that knowledge. If the world had not been at war in the 1940s, the discovery of the relationship between matter and energy at the level of the nucleus might not have led, or at least not as rapidly, to the development of nuclear weapons. Yet many scientists worked on the atomic bomb project with little sense of its enormous consequences. Some, like the physicist, Robert J. Oppenheimer, were politically naive, assuming that the generals in charge of the project would consult the physicists about how and whether the bomb would be used. Others lost sight of the implications of nuclear energy in the sheer pleasure of solving a set of technical problems. Now that the Cold War is over without a single exchange of nuclear bombs, debates over the morality of science have taken another turn: who decides what science will be done, who pays, who benefits, and who participates?

The nuclear disaster at Chernobyl in 1986 and the near-disaster when a U.S. nuclear reactor went "critical" at Three-Mile Island a few years before, alerted the world to how vulnerable we all are to windswept radioactive fallout and other environmental insults. Acid rain, as the world has discovered, knows no national boundaries. Global warming, if it occurs, will affect us all alike. Although environmental issues are now being addressed primarily in the developed nations, they must eventually be considered in all nations, as people realize that natural resources are not infinite and that every decision involves a trade-off. If they use their knowledge constructively, scientists can help us make the best of the resources we have.

Nearly 20 years ago, when the possibility of biologically engineering new species first came into view, a group of biologists at the National Institutes of Health came together to set policy for research in this area. Their goal was not to stifle research, but to manage it. They understood that scientists must take responsibility for the discoveries they make and for the uses to which these discoveries are put. As you make your way through your training in science, we hope you will think about the social responsibility of science, and the duties it entails: to seek the truth; to communicate knowledge to all who need it, in language they can understand, and to use expertise for the betterment of humanity.

Further Reading

General Interest

Gould, Stephen Jay. *The Mismeasure of Man.* New York: W. W. Norton & Co., 1981.
Hawking, Stephen W. *A Brief History of Time: From the Big Bang to Black Holes.* Toronto: Bantam Books, 1988.
Kuhn, Thomas S. *The Structure of Scientific Revolutions.* Chicago: The University of Chicago Press, 1970.
Snow, C. P. *The Two Cultures.* London: Cambridge University Press, 1964.
The Astronomers. A series of six videotapes from PBS Home Video. Pacific Arts Video Publishing, 1991.

The History of Science and Scientific Discovery

Cohen, I. Bernard. *The Birth of a New Physics.* New York: W. W. Norton & Co., 1985.
Gleick, James. *Chaos: Making a New Science.* New York: Viking Press, 1987.
Hazen, Robert M. *The Breakthrough: The Race for the Superconductor.* New York: Ballantine Books, 1988.
Keller, Evelyn Fox. *A Feeling for the Organism: The Life and Work of Barbara McClintock.* New York: W. H. Freeman & Co., 1983.
Levi, Primo (translated by Raymond Rosenthal). *The Periodic Table.* New York: Schocken Books, 1984.

Sayre, Anne. *Rosalind Franklin & DNA.* New York: W. W. Norton & Co., 1975.

Watson, James D. The Double Helix: *A Personal Account of the Discovery of the Structure of DNA.* New York: The New American Library, Inc., 1968.

BIOGRAPHY

Biographies and autobiographies of some modern scientists are published in the Alfred P. Sloan Foundation Series by Basic Books, Inc., New York.

Alvarez, Luis W. *Alvarez: Adventures of a Physicist.*

Bruner, Jerome. *In Search of Mind.*

Casimer, Hendrick. *Haphazard Reality.*

Dyson, Freeman. *Disturbing the Universe.*

Levi-Montalcini, Rita. *In Praise of Imperfection: My Life and Work.*

Luria, S. E. *A Slot Machine, A Broken Test Tube.*

Ridgen, John. *Rabi: Scientist and Citizen.*

York, Herbert F. *Making Weapons, Talking Peace.*

MATHEMATICAL INTRODUCTION

Paulos, John Allen. *Innumeracy: Mathematical Illiteracy and Its Consequences.* New York: Hill & Wang, 1988.

Thompson, Silvanus P. *Calculus Made Easy: Being a Very-Simplest Introduction to Those Beautiful Methods of Reckoning Which Are Generally Called by the Terrifying Names of the Differential Calculus and the Integral Calculus.* New York: St. Martin's Press, 1946.

Index